# HARTLAND® HORSES

# NEW MODEL HORSES SINCE 2000

Gail Fitch

4880 Lower Valley Road • Atglen, PA 19310

# DEDICATION

This book is dedicated to my parents, Betty and Lawrence Fitch, who celebrated their 62nd wedding anniversary on September 10, 2011.

*Photo by Johannes Studio, Milwaukee, Wisconsin, 1949.*



**Other Titles by Gail Fitch**
*Hartland Horsemen* ISBN 0-7643-0947-1 $29.95
*Hartland Horses & Dogs* ISBN 0-7643-1268-5 $29.95



Library of Congress Control Number: 2012935009



Designed by Mark David Bowyer
Type set in Anna / Century Schoolbook

ISBN: 978-0-7643-4028-4
Printed in China

Schiffer Books are available at special discounts for bulk purchases for sales promotions or premiums. Special editions, including personalized covers, corporate imprints, and excerpts can be created in large quantities for special needs. For more information contact the publisher:

Published by Schiffer Publishing Ltd.
4880 Lower Valley Road
Atglen, PA 19310
Phone: (610) 593-1777; Fax: (610) 593-2002
E-mail: Info@schifferbooks.com

For the largest selection of fine reference books on this and related subjects, please visit our website at
**www.schifferbooks.com**
We are always looking for people to write books on new and related subjects. If you have an idea for a book, please contact us at
proposals@schifferbooks.com

This book may be purchased from the publisher.
Include $5.00 for shipping.
Please try your bookstore first.
You may write for a free catalog.

In Europe, Schiffer books are distributed by
Bushwood Books
6 Marksbury Ave.
Kew Gardens
Surrey TW9 4JF England
Phone: 44 (0) 20 8392 8585; Fax: 44 (0) 20 8392 9876
E-mail: info@bushwoodbooks.co.uk
Website: www.bushwoodbooks.co.uk

# CONTENTS



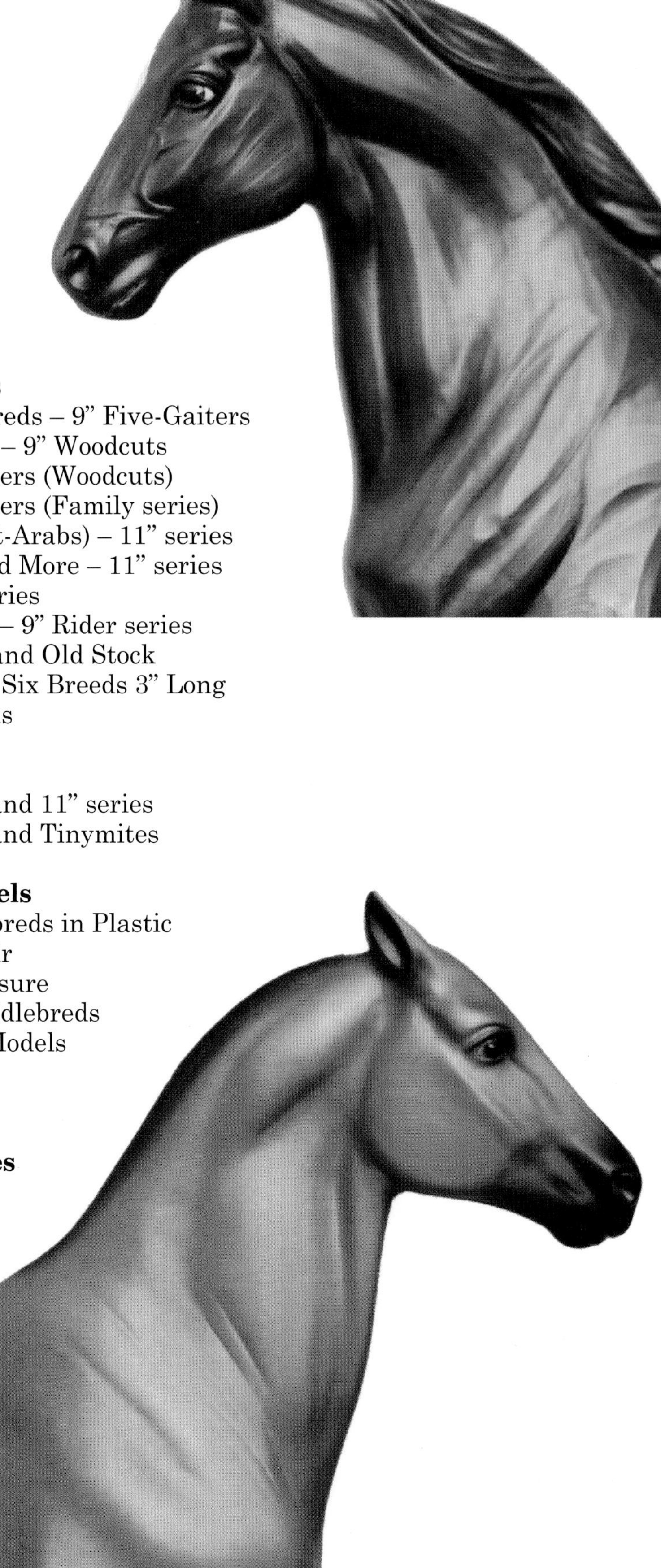

# ACKNOWLEDGMENTS

More than two dozen people were kind enough to share their model horse collections through photographs. I am grateful to: Anni Koziol, April Powell, B & K, Chelle Fulk (and photographer Andrew D. Culhane), Deana Sprague, Dee Gwilt, Deirdre Price, Denise Brubaker, Denise Hauck, Dianne S. Teachworth, Doris Hunter, Elaine Boardway, Eleanor Harvey, Judy O'Bannon, Lori Ogozalek, Lynn Isenbarger, Marla Phillips, Melanie Teller, Melissa Clegg, Nerissa Pospychala, Pamela Pramuka, Pat Noble, Robert Ezerski, Robyn Porter, Sylvia Hand, Tammy Nguyen, and Victoria Zanutto.

Also helpful were: Cassie Hayes, Donna Anderson, Ellen Maceko, Jan Kreischer, Jennifer Pomerance, Kat Jennings, Laura Pervier, Lauri Barnwell, Leslie Reynolds, Linda Walter, Lydia Fey, Lyn Arnold, Sandra Truitt, Sharon Elosser, and Suzy Meathrell. Others are named within. I am also grateful for many collector friends not mentioned here.

Collector Cecile Bellmer served as editorial consultant. Non-collector friends and family who helped, in various ways, include: Roger C., artist Michael Mueller, and my father, Lawrence Fitch, who suggested an improvement to the first edition.

For first edition printing, I am grateful to Milwaukee's Digicopy on Van Buren Street and FedEx Office (formerly Kinko's) on Farwell Avenue, especially to Daniel and Kevin. For photo developing and printing, I have Walgreens on Brady Street to thank, and especially: Andrew, Yvonne, Daisey, Daniel, Eric, Lian, and Jarrett.

One of the first products of the Hartland company that formed in 2000 was this 7" Tennessee Walker Family, 684-01, from March 2001. *Courtesy of Deirdre Price.*

# PREFACE

For almost three decades, Gail Fitch's books on Hartland have celebrated a fine product line appreciated by horse lovers and collectors, defined the subject, and uncovered, gathered, reconstructed, recorded, and preserved history that would otherwise have been lost.

Gail joined the model horse collecting community in 1975 by subscribing to model horse journals. Judging by the journals, little was known about Hartland, and information (such as old company catalogs) on Hartland was harder to find than for any other major American brand of model horse. Gail decided to apply her experience in research, writing, and publishing to her hobby field, model horses, by doing a book on Hartland. After all, the company had, before it closed in Wisconsin in 1978, been located less than 30 miles from her home. Starting research from scratch, she began locating and interviewing Hartland officials in 1980. It became a full-time passion.

For sixteen years (1983-1998), she self-published books in full color and sold them at a nonprofit price. The first book, *Hartland Horses and Riders* covered the model horses, horse-and-rider sets, and company history and artistry. In 1983, it was 72 pages; the sixth edition, in 1995, was 239 pages. She printed up to three times per year to keep the book available almost continuously.

Gail also published eighteen monthly issues of a Hartland newsletter in 1994-1996. Its subscribers, like the book buyers, included not only model horse hobbyists, but also rider-set collectors, and others outside the model horse hobby.

Finally, a publisher was interested in the subject. Gail's best work on the pre-2000 Hartland models is in two earlier books from Schiffer Publishing: *Hartland Horses and Dogs* (2000), on model horses and the Hartland story; and *Hartland Horsemen* (1999) on the horse-and-rider sets. This latest book, on the post-2000 Hartland models, continues Gail's thirty-year labor of love.

A gaited horse in gold champagne (palomino with pink skin and amber eyes) waited quietly in a stall at the Midwest Horse Fair in Madison, Wisconsin, April 1991.

# INTRODUCTION

Five-Gaited Saddlebreds stepped past, keeping perfect time. Newly corralled Mustang stallions reared and wheeled in mock fights over imagined mares. Tennessee Walkers nodded along. As handlers led them, mares kept an eye on their foals.

Neighs rang out from the Arabian stallions standing under the shade trees. Quarter Horses, Appaloosas, and Paints snorted and pranced in the paddock. Polo Ponies, nimble and alert, dashed about at their riders' commands. In the distance, approaching horses made a moving quilt of ebony, copper, and gray; browns, reds, and yellows.

There were arched necks, dappled sides, flared nostrils, and soulful eyes. Everywhere, coats gleamed in the sun.

Children were awed. Old timers sighed. The crowd thronged. It was a splendid horse fair.

A new Hartland company, Hartland Collectibles, L.L.C. formed in 2000, and began issuing model horses in 2001. This is the story of those horses, in three parts:

## PART 1 - PRODUCTION MODELS

Production models were runs of models that it was possible to buy. They include regular runs and special runs. Four of the models could only be purchased by the winners of a drawing, but most production models could be bought outright. The production models are subdivided into chapters by mold or shape, for example: Mustangs, Polo Ponies, etc.

## PART 2 - GIFT RUN MODELS

Gift models were tiny runs of models that could not be purchased; instead, they were only given as gifts or as contest prizes whose winners paid nothing for them. The gift runs varied from three to 20-something pieces, but usually numbered fewer than 10.

## PART 3 - TEST (UNIQUE) MODELS

Part 3 is for the unique ("one-of-a-kind") models. Since 2000, the Hartland "test models" were unique models (not runs). It describes all of the known test or unique models and illustrates a good share of them. This part also includes some miscellaneous models that fall short of being unique, but are neither production models, nor gift-run models.

**Values.** The introduction to Part 1, Production Models, discusses condition assessment and value rating categories and their abbreviations. The introductions to Part 2 and Part 3 discuss values for the gift-run and test models, respectively.

A Quarter Horse in dappled buckskin (902-11), looks down on a polo game at the Santa Barbara (California) Polo and Racquet Club, June 1989.

## Horse and Dog Shapes

Let's begin with an overview of the 24 horse and dog shapes produced (and sold or donated by Hartland) since 2000. The 23 horses and one dog are shown in Photos A-F. The examples pictured are all from the Hartland 2000 era, except for rare shapes. After 2000, the 9" Thoroughbred appeared only as gift runs and tests; the example in photo A is from the 1960s. The four shapes in photo F were used after 2000 for only a handful of test pieces, so the colors shown are from the 1980s-1990s.

Here are some observations about the shapes. The Polo Pony, photo B, has leg and tail wraps molded on. Molded-on tack is rare in the breed series, but Hartland's rider-series horses all came with reins: either loose or molded-on. The 11" series Quarter Horse, photo C, has its head tucked, but the Tinymite with head tucked is the Morgan. The 9" Tennessee Walker, photo D, is the version with the mane on both sides.

The German Shepherd (photo E) is a model of Roy Rogers' dog, Bullet. The standing / walking Chubby horse with "wrong tail" got it from a different rider-series horse, due to a factory assembly error. The Chubby horses are the variation with molded-on bridle. Two Semi-Rearing horse types were used after 2000: the one in photo E has "mane up and plain tail," and the one in photo F has "mane down."

**A**
The shapes used since 2000 include: the 9" series Thoroughbred *(bottom left)*, 9" series Five-Gaited American Saddlebred, and Tinymite Belgian *(top)*.

**B**
These shapes are the 9" series Polo Pony *(bottom left)*, 11" series Arabian Stallion, and Tinymite Arabian *(top)*.

**C**
More shapes used were, *bottom:* 9" series Mustang woodcut (rearing); and 11" series Quarter Horse; *top, from left,* the Tinymite Thoroughbred, Quarter Horse, and Morgan.

**D**
Tennessee Walkers ("TWs") produced after 2000 are, *upper:* 7" TW Foal, 7" TW Mare, and Tinymite TW; *bottom:* 7" TW Stallion, and 9" series Tennessee Walker.

**E**
Shapes used from the 9" rider series are, *top:* the German Shepherd, and Chubby horse (wrong tail); *bottom:* Chubby (correct tail), and Semi-Rearing horse (mane up, plain tail).

**F**
Shapes painted after 2000 for tests only are, *top:* the 9" Weanling Foal and 9" Semi-Rearing horse (mane down); *bottom:* the "Lady Jewel and Jade" 11" series Arabian Mare and Foal.

| ANIMAL SHAPES AND SIZES | | | | | | | |
|---|---|---|---|---|---|---|---|
| **Mold Name** | **Photo** | **Tall** | **Long** | **Mold Name** | **Photo** | **Tall** | **Long** |
| 11" Arabian Stallion | B | 9.75" | 11.5" | 11" Lady Jewel (T) | F | 8.25" | 10.25" |
| 11" Quarter Horse | C | 7.75" | 11.25" | 11" Jade (Foal) (T) | F | 6.5" | 6" |
| 9" Five-Gaiter | A | 8.25" | 9.5" | 9" Thoroughbred (G & T) | A | 7.75" | 10.25" |
| 9" Polo Pony | B | 8.25" | 9" | 9" Weanling Foal (T) | F | 6" | 6" |
| 9" Mustang Woodcut | C | 9.5" | 8.75" | 9" Semi-Rearing, mane up | E | 8.5" | 9" |
| 9" Tennessee Walker | D | 8.25" | 9.5" | 9" Semi-Rearing, mane down (T) | F | 8.5" | 8.5" |
| 7" Tennessee Walker Stallion | D | 6.5" | 8" | 9" Chubby w/ correct tail | E | 8.25" | 8" |
| 7" Tennessee Walker Mare | D | 6" | 7.25" | 9" Chubby w/ wrong tail | E | 8.25" | 8.25" |
| 7" Tennessee Walker Foal | D | 5" | 4.25" | 9" series German Shepherd Dog | E | 4" | 6.5" |
| Tinymite Arabian | B | 2.5" | 3" | Tinymite Quarter Horse | C | 2.5" | 3" |
| Tinymite Belgian Draft Horse | A | 2.25" | 3.25" | Tinymite Tennessee Walker | D | 2.75" | 3" |
| Tinymite Morgan (head tucked) | C | 2.25" | 3" | Tinymite Thoroughbred | C | 2.25" | 3" |

**Animal Shapes and Sizes** — These 24 animal shapes (23 horses and one dog) encompass the 2000-2007 plastic, production run and gift run models and the uniquely painted test models that were sold. The shapes are shown in photos A-F. Four sizes (scales) are represented. Measurements were rounded up to the nearest one-quarter inch. The series names – 11", 9", etc. – reflect the lengths of the models. In the "Mold Name" column, the four shapes marked with "(T)" were used only for test colors. The 9" Thoroughbred ("G & T") was used for gift runs and test pieces, but not for production runs.

## HOW IT BEGAN

Hartland model horses were first made in Hartland, Wisconsin, in the 1940s, and were produced there by Hartland Plastics, Inc., until 1969. Then, the molds and brand name traveled twice to new owners before being sold again in 2000.

The new model horses released since 2000 actually represent only about one-third – but a nice third – of all the different Hartland horse shapes that have been created since the 1940s. The molds used since 2000 are from 1958-1967, except "Lady Jewel" and "Jade" which are from 1988.

The dates and other information on all of the horse shapes from the 1940s through 1990s were extensively researched and can be found in my other two books. *Hartland Horsemen* (1999) encompasses the horse-and-rider sets and all rider-series horses, even when they were sold without a rider. The subject of *Hartland Horses and Dogs* (2000) is the breed-series horses, and that book includes the very detailed (and illustrated) history of Hartland.

The pre-1980s Hartland sculptors were Roger Williams and Alvar Backstrand, truly great artists. "Lady Jewel" and "Jade" were sculpted by Kathleen Moody for Steven Manufacturing Company, producer of Hartland horses from 1983-1994. Since then, Kathleen Moody has sculpted models for Breyer Animal Creations. Steven Manufacturing stored the Hartland molds from 1994 until the sale to Hartland Collectibles, L.L.C., in 2000. *Hartland Horses and Dogs* was published in December 2000. It had barely arrived in stores when Hartland horses returned in February 2001!

## WHAT TO LOOK FOR IN THE MODELS, AND HOW THIS BOOK DESCRIBES THEM

Information on plastics, earlier Hartland companies, horse color as it applies to models, and other terms and definitions are found on pages 6-7 in *Hartland Horses and Dogs* and pages 6-7 in *Hartland Horsemen*, so I won't repeat all of that here, but a few things should be mentioned.

**The Hartland Company.** Hartland Collectibles, L.L.C., which formed in 2000, created two divisions: one for sports statues, which were its primary interest, and a "western" division for model horses and horse-and-rider sets. The western half often called itself: Hartland Horses. Since there have been several Hartland companies with similar names, I will sometimes refer to the company that formed in 2000 by a nickname, Hartland 2000.

**Shapes and Models.** An example of a shape (or mold) is the 9" Five-Gaiter. A model, in the collective sense, is a particular shape in a particular color. For example, the Palomino Five-Gaiter is a model. Two horses of the same shape, but different colors, are two different models. A "mold" can also refer to the equipment used to manufacture the plastic pieces. The Hartland models were molded in parts and then glued together by the factory.

**Plastic Type.** The company announced that the plastic horses would be cellulose acetate plastic. That is the more desirable of the two types of plastic most frequently used for model horses of all brands over the last sixty-plus years. The less desirable alternative, styrene, is more brittle.

**Model Horse Color.** Here's a quick summary of some common (real) horse colors. It will be useful for understanding the models. A yellow or beige horse with white mane and tail is usually a **palomino**. A yellow or beige horse with black (or brown) mane and tail is usually a **buckskin** or **dun**. A horse with a brown, reddish, or tan body and black mane and tail is usually a **bay**. Horses with a brown, reddish, or tan body and mane and tail that are lighter, matching, or slightly darker are usually **chestnut** (or **sorrel**). Just being able to tell bay from chestnut will go a long way – with models and with real horses.

Among spotted horses, **pintos** are the ones with white areas usually the size of dinner plates or larger. Spots more like the size of a 50-cent piece, and relatively solid, are found on **appaloosas**. Spots the size of a 50-cent piece, but that are just loose clusters of hairs, not solid or neatly round, and that are white, are called dapples, as seen on **dapple greys**.

Horses that are a mix of white and colored (nonwhite) hairs over most of their body, and that progress in their lifetime from all colored to nearly all white, are called **grey**, spelled with an "e." A grey will usually have a lot of white on its head, and often, in its mane and tail. That is the easiest way to tell greys apart from roans. Typical **roans** have white and non-white hairs mixed on the body, but their head and points – mane, tail, and lower legs – are the non-white color, except for any white leg or face markings that might happen to be present. In all horse colors, white markings on the face and lower legs are incidental to the horse's color, not a defining factor.

This book includes models in bay roan, chestnut roan, and blue roan. A real blue roan horse has black and white hairs mixed evenly all over its body, black hair on at least most of its head, a solidly black mane and tail, and black on the lower legs (except, perhaps, for white stockings). In model horses, this color is often represented by gray paint on the horse's body to simulate the mix of white and black hairs. In this book, models described as "blue roan" are not literally blue; they are gray.

Getting back to grey, in model horses, a "white-grey" has a white body without shading or paint specks. It might sound silly, but a "gray-grey" model is one that represents the grey horse color by gray paint on its body. Model horses come in fantasy colors, too. Real-horse color terms are the starting point for describing the color of model horses, but it doesn't end there.

**Dapples.** Dapples on models are small, roundish spots within a matrix of darker shading or a darker base color. The dappling is relatively uniform, but not exactly the same, on each example of the same model. On a real, dapple-grey horse, the dapples are white and about the size of a quarter or 50-cent piece. Some real horses in colors other than grey, show subtle, dark (reverse) dappling. The models made since 2000 include fantasy colors, and so some have silver dapples, instead of white dapples, or in various other ways, do not follow the rules of horse color. (Too much realism can be boring!)

**Points.** I use the term "points" to refer to the mane, tail, and lower legs, collectively. That does not include the hooves. On a real horse, when the color on the lower legs is darker than the body in general, the darker color typically extends up at least part way onto the knees and hocks. Hartland 2000 often carried the dark lower-leg color onto the knees and hocks so thoroughly that I mention the knees and hocks separately. Thus, for emphasis, some models are described as having, "black points, knees, and hocks."

**Dark Details.** Most of the post-2000 models have dark shading – either black, dark gray, or brown – free-hand airbrushed on the muzzle, inside the nostrils and ears, and in the groin area. The model descriptions in this book often mention the muzzle color, but usually omit the smaller or more hidden areas. You can assume that the dark shading is there.

**Pale Details.** This book usually describes the light-colored hooves as "peach," but sometimes calls them "pink." The terms are usually interchangeable. The typical hoof color looks a little closer to pale orange than pale pink, but in any event, has a hint of beige.

On the face, some of the white markings turn to peach or pink between the nostrils. Peach or pink is also found in the corners of the eyes on some models.

**Details** – A white face marking painted on over the base color was typical of the detailed effort by Hartland 2000. This example, a gift-run Bay Thoroughbred, 873-BG, has a star. *Courtesy of Elaine Boardway.*

**Finishes and Markings.** Most of the Hartland 2000 models have a satin or semi-gloss finish while others are matte. Some have a metallic or "pearled" sheen. The metallic gold and metallic copper models have tiny, reflective flecks of those respective colors in their paint.

Found more often, though, is a silver-pearled effect that gives a reflective sheen to body colors ranging from black to palomino. Even more striking are the brilliant, pearled-white markings on some of the models. Still, other models appear to have a pearled undercoat. In some instances, a pearled coat is only visible on the lower legs, as white stockings.

Some models have areas of bare plastic, such as on the lower legs (as white stockings). The bare plastic is white and glossy. In most instances, though, white areas were painted white: either normal white or pearled white. Normal white paint is "flat" (not reflective). White face markings were nearly always painted on, with neatly defined edges.

**Variations within Runs.** Minor details in how the models were painted may vary from piece to piece, but there was generally a high level of consistency. A few production models from 2001 did vary significantly and intentionally. The flecked grey Polo Ponies, black-gray pinto Mustangs, black pearled Arabians, and "Silver Sultan" (the dappled bay Arabian) are found in noticeably lighter and darker versions. Also, in 2006, each Red-Chestnut Overo Pinto Polo Pony was deliberately painted with variation in the pinto markings.

**Glamour** – The white areas on the Palomino Five-Gaited 9" Saddlebred, 881-01, are pearled white, and its body photographed as glowing gold. *Courtesy of Deirdre Price.*

**Elegance** – This photo draws out the reddish tone in the Pearled Palomino Five-Gaited 9" Saddlebred, 881-01. It's a graceful, elegant model. *Courtesy of Marla Phillips.*

## STREAMLINING THE MODEL DESCRIPTIONS

**Color Labels.** For clarity, and in order to distinguish these latest models from earlier Hartland models, I've slightly rephrased some of the color names the company used, but I also mention the company's color label for the model. For listings, my notation style is to usually place the obvious feature(s) of a model's color first, and then mention less-apparent features, such as dapples or a pearled finish, which on some models may be almost undetectable in a photo. For example: "Dark Bay (Pearled)" or Palomino (Dappled) might appear in a heading or in a table. However, when used in a sentence, the description will follow the normal spoken word order; as in, "The Pearled Dark Bay ... has four socks" or "The Dappled Palomino ...has a blaze." Collectors are likely to shorten these labels to "Pearl Dark Bay" and "Dapple Palomino," and I will sometimes do that, too.

During the first year, when Hartland 2000 used the word "silver" in a model's color label, it meant the model had a pearled (silver-pearled) finish. After I published "pearled" in place of "silver," the company followed my lead. After all, a black model with a silver-pearled finish looks black, not silver, so "Black (Pearled)," "Black Pearl," or "Pearl Black" are better labels than "black silver." "Silver" also connotes a group of dilute real-horse colors; when used in a model description, it could pose an additional dimension of potential confusion.

**Rose Grey.** Among real horses, many greys display a mix of only black and white hairs even when, genetically, their base color is bay, chestnut, or another color other than black. Greys with "warm colored" hairs (brown, tan, yellowish, or reddish hairs) are less common, and are often described as "rose grey." The term "rose grey" was originally applied to horses whose coat looked pinkish from a distance due to being a mix of white and light chestnut or light bay hairs. Some of those horses were greys, and some were actually red roans with an additional genetic trait, "frosty," that resulted in white hairs mixing into the head and points. (A pinkish grey model is on page 9 in *Hartland Horses and Dogs* and a roan-type rose grey is illustrated in *The Color of Horses* by Dr. Ben K. Green, 1974.) However, "rose grey" is usually used more broadly to include greys with any shade of "warm colored" hairs, and during phases ranging from when the horse is nearly all colored to nearly all white, especially if the mane and tail have retained color. (Greys that are more than about 95% white with reddish or brown hairs randomly scattered, not clustered to encircle dapples, and with white mane and tail, are called flea-bit grey.) Hartland 2000 used "rose grey" so broadly that I felt more definition was needed.

**Series / Size Names.** By the mid-1960s, the original Hartland company, Hartland Plastics, Inc., issued a series of horses in five scales, and labeled the series as: Tinymites (3"), 5", 7", 9", and 11". It sometimes called the 11" series the "Regal" series. Hartland 2000 called the 11" series the "Regal" series, the 9" series the "Heritage" series, and the 7" series the "Legacy" series, except that 9" and 11" models painted by the company's designer were called the "Noble Horse" series. For this book, I'm sticking with the traditional terminology: 11", 9", and 7". It's simpler.

While the horses are often at least one-half inch less in height than the series names imply, their lengths live up to the billing. For example, the 11" series Arabians molded after 2000 measure 9.75" tall, but are 11.5" long. I think Hartland Plastics must have intended the series names to represent the lengths of the models, not the heights.

In my Hartland books from 1983-2010, I listed only the heights of the models. I should have guessed sooner. The original Hartland company was nothing if not honest: There was "truth in advertising." It certainly is poetic that Hartland made horse series in 3", 5", 7", 9", and 11" sizes, and those were accurate labels, too.

**Determining Length.** The length of a model horse is measured horizontally, not diagonally. It is the horizontal distance between a line dropped from the front-most point of the horse (often, the nose) to a line dropped from the point farthest back (the tail or a hoof extended backward). Here's a fun way to measure the lengths. You'll need two or three feet of table space, two boxes about 10" tall, and a 12" ruler. Put a box on the table. Put the horse on the table, perpendicular to the box, with its nose (or whatever is farthest forward, which could be a front hoof, the ears, or the forelock) touching the box. Lay the ruler on the table, parallel to the horse, and with its one-inch end touching the box. Then, place the second box against the farthest-back point of the horse, and alongside and behind the ruler. Then, read the horse's length where the second box touches the ruler.

Measure model horse length horizontally, from a "wall" (box) touching the farthest point forward to a "wall" touching the farthest point back. On the 7" Tennessee Walker Stallion, those points are his nose and lower tail. Use boxes or other "wall" objects that stand straight and don't have bowed sides. This horse is 8" long, but the ruler reads 8.25" because the box wasn't plumb.

**Catalog Numbers.** The catalog numbers that Hartland 2000 used are a three-digit "mold" number followed by a hyphen and numbers or letters to designate the "color" of the model. After the first few years, the company left some of the numbers incomplete (nothing after the hyphen), so I completed them with a letter code, much like the company, itself, sometimes did.

The numbering on the production colors of a particular shape does not necessarily match the release order, and some numbers were skipped entirely. To use a made-up example, let's say that Arabian color #6 was released before Arabian color #4, and there was no Arabian color #5. I think part of the reason for the skipped numbers is that each test (unique) model was numbered, but many test colors were not followed by a corresponding production run.

## Additional Definitions and Abbreviations

Some handy abbreviations I use in this book are:

- ASB = American Saddlebred.
- Equine Affaire is the name of a particular real-horse convention with a trade show and demonstrations. In recent years, it has been held annually in Ohio, Massachusetts, and California.
- Horse-Power Graphics, Inc. = A business owned by Sheryl Leisure, who was also horse director for Hartland 2000. (After the first mention in a chapter, the name may be abbreviated as "HPG," or "HPGraphics," or "Horse-Power Graphics," without the "Inc."
- Jamboree = West Coast Model Horse Collector's Jamboree, an event that was held annually in California by Sheryl Leisure / HPGraphics. Many Hartland special runs (and some test colors) were funneled through the Jamboree.
- JSC = Jamboree Summer Challenge. That was the larger of two model horse shows held at the Jamboree. The JSC was the open show, as opposed to the novice show. The JSC kept its name whether it was held in June, August, or October.
- QH = Quarter Horse.
- r – Used in information tables, "r" signifies a regular run
- s. = Series, as in 7" series, 9" series, 11" series.
- SR = special run
- TW or TWH = Tennessee Walker or Tennessee Walking Horse
- w/ = with

## Numbering (Hand-Written)

The factory put handwritten information on the underside of some models.

If a model has nothing written on it, you can assume that it is from a production run that was not a numbered edition.

**Production Models.** Some of the special-run production models were hand-numbered (in pen) on their underside. They were club models, holiday models, and the "Noble Horse Series." They are typically marked with a three-digit number, such as: #039.

**Gift Run Models.** Most of the gift-run horses were Hoof Pick Contest prizes, and came with a paper certificate (a letter). An example of gift-run notation is: '04 Hoofpick 1 of 3. That was written in fine-point, black marker pen on the underside of a 2004 gift-run model.

**Test (Unique) Models.** A six-or-eight-digit number hand-written on the underside is the sign of a test (unique) model. Often, but not always, the letter "T" or the word "Test" will appear, also. An example of the style of test model notation is: 881-010T.

(The production models have a catalog number like 881-10, but the number does not appear on the models, themselves.)

## Model Information and Sources

For the production models, the historical information in this book includes month and year of debut, catalog item number, original price, and edition size. This book also tracks the changes in price and edition size, and mentions when each model was close to being sold out.

I ordered the new Hartland models from the company promptly, and usually count the month on the receipt enclosed with the model as the month of debut. The models' prices came from the receipts and the company's catalog sheets, price lists, etc. The catalog numbers came from the company's newsletter; they were never published at the company website. The edition sizes and quantity remaining were mentioned at the company website from mid-2000 through mid-2008. At least a few times per year, I took notes on the quantity of models remaining.

The articles I wrote about Hartland, from late 2000 on, made the work of preparing this book much easier. In November 2000, I started writing about Hartland 2000 at my website, www.hartlandhorsemen.com, and in December 2000, I published my first issue of *Hartland Model Equestrian News*. Each issue after that, starting with March/April 2001, included a table listing the latest model releases and/or the number of models remaining from each run. I also wrote about the new models in articles for *Horsing Around Magazine*, England.

The first issue of the company newsletter to include a list of models was the Winter 2003 issue, which arrived in April 2003. Once a year, the company added to its list. I also referred to the Hartland company's advertisements and news releases in 2000 to 2005 in *The Hobby Horse News* and *Horsing Around Magazine*, company announcements by email and mail, and a few at model horse groups at www.yahoo.com.

Other sources included my notes on the company's eBay auctions of test models and eBay store sales of production models. The Hartland company eBay store, hartland_horses, was no longer used by early August 2008; it closed in 2009. Since 2008, Hartland's remaining models have been sold in the horse_power_graphics_inc eBay store. Those listings sometimes mentioned the run size and quantity remaining.

Additional information on the new Hartland models came from phone calls or correspondence with Hartland owners or staff, mailings from HPGraphics, and notes from viewing Hartland special runs and test colors for sale at the HPGraphics / Jamboree website, www.modelhorsejamboree.com, and from the Jamboree printed programs.

Last, but not least, correspondence with fellow Hartland enthusiasts about their test color and gift run models, and variations in some of the production models, was indispensable. Other sources are noted in the text.

## Brand Marks

Hartland brands over the years are covered on page 10 in *Hartland Horses and Dogs*, and, in fact, I am the person who (in the 1990s) found decades-old documentation of the brand marks. Most horse collectors can recognize a Hartland horse by its shape and size, and do not need to see a brand mark to know that it is a Hartland. However, I'll summarize the brand marks appearing on the model shapes produced since 2000.

These 11 models never had a Hartland brand mark before, and still didn't after 2000: the six Tinymite breeds, the 7" Tennessee Walkers (all three family members), the 9" Mustang woodcut (current version), and the 9" Tennessee Walker with two-sided mane.

For the animals marked with the word "Hartland," their brand mark is unchanged since the last time their molds were used. (The brands are raised lettering.)

- Marked "© Hartland" on the right hind leg are: the 9" Five-Gaiter, 9" Polo Pony, 9" Semi-Rearing horse with mane up and plain tail, 9" Weanling Foal, 9" Thoroughbred, and Semi-Rearing horse with mane down.
- Marked "© Hartland Plastics, Inc." are the 9" Chubby with bridle molded on (marked on the right hind leg), and the German Shepherd Dog (marked on the left hind leg).
- The "Lady Jewel" and "Jade" models are marked, "Hartland/Steven" and "Moody 88" in engraved letters.

Two shapes had their mold mark altered since they were last used in the 1990s: the 11" Arabian and 11" Quarter Horse. In the 1980s-1990s, those horses were brand-marked: "Hartland / Steven." Sometime between fall of 2000 and June of 2001, the brand was removed (by filling it in and covering it over on the tooling). Most examples of those two shapes released since 2000 now have just a smoothed area on the inner, upper left hind leg where the brand had been.

However, the purchase from Steven in fall of 2000 included some unpainted horses that had been molded in the 1990s. Thus, some of the 11" Arabians and Quarter Horses have the "Hartland / Steven" brand mark. One of my 11" Quarter Horses has the "Hartland/Steven" brand mark. One or more of yours might have it, too.

## On Gender in Hartland Horses

In the 1960s, Hartland catalogs labeled some models as stallions or mares, but for most of the models, the catalogs did not specify their gender. The horses that the catalogs called a "stallion," had gelding anatomy. The models the catalogs called a "mare" had a smooth underside (groin area).

The catalog did not describe any models as geldings. However, I think one of the smooth-groined horses whose gender was not specified in the 1960s catalogs – the 11" series Five-Gaiter – was a gelding. I think it was closely modeled after a Saddlebred gelding pictured in a horse book of the time. (See page 155 in *Hartland Horses and Dogs*.) The 11" Five-Gaiter was not used by Hartland 2000.

Also, all 12 Hartland foal shapes from the 1960s have smooth groins. I doubt that the sculptors would have intended all of the babies to be specifically female. So, my conclusion is that the Hartland sculptors' convention was that gelding anatomy appeared on stallions, and everyone else – mares, geldings, filly foals, and colt foals, got a smooth underside.

Using gelding anatomy as the mark of a stallion, these seven shapes used since 2000 are stallions: the 11" Arabian, 9" Mustang, 9" Tennessee Walker, three of the Tinymites – the Arabian, Quarter Horse, and Morgan – and of course, the 7" Tennessee Walker Stallion. I think the other adults – those with smooth groins, and whose genders were unspecified in the 1960s catalogs – could be either mares or geldings. Take your pick.

Hartland's minimalist approach to gender anatomy allows some latitude. Collectors should feel free to assign whatever gender they think their Hartland model resembles by its head shape and carriage, facial expression, general outline, overall musculature, etc. That is similar to assigning a breed to your model.

In this book, I sometimes refer to the smooth-groined models as "him," but other times, as "her." Owners of the models vary on this point, too. (Talk about gender inequity: In the world of factory-made model horses, stallions still outnumber mares, but geldings are, by far, the most overlooked gender. Historically, Hartland and Hagen-Renaker have offered a better balance of the genders than Breyer has.)

## Hartland Horses Not Tethered

If model horses were judged like figure skaters, Breyer (and Peter Stone) would score well on "technical merits" – the "realism" of their anatomy and "conformation" – but Hartland would wow the judges with "artistic impression." Which skater would you not want to miss?

The Peter Stone Company, which began in 1995, has pushed back the limits on dramatic color designs while also producing a wide-range of real-horse coat colors. In the 1960s, Breyer made a few decorator colors in small quantities, and over decades, slowly broadened its palette beyond the ordinary. For Hartland, its 1960s copper-orange horses were best sellers in a line that included glossy red, metallic blue, and pearl white horses alongside bays, chestnuts, and palominos. The 1980s and 1990s Hartland companies added some of the most detailed and realistic colors seen on model horses of their time, and Hartland 2000 added yet more attractive colors: some realistic, some fanciful.

Art is as much about beauty as realism. In conformation and color, Hartland horses have never been tethered to reality. So, leave your "Breyer thinking" at the door. Prepare to soar beyond the familiar. Our magic mounts await.

**Spirit –** The Hartland Mustangs are free spirits in a powerful, rearing pose. This example from 2001 is the Chocolate Bay Mustang woodcut, 886-01.

# PART 1
# PRODUCTION MODELS

The production models are the regular-run and special-run items that were for sale. That excludes gift-run models, which were not for sale, and test (unique) models. The production models are divided into separate chapters – generally, one for each horse shape or closely related group of shapes. Then, Chapter 11 goes into more detail on the production items.

**Values for Production Models.** First of all, the values in this book apply only to the specific models illustrated. Production-run horses of the same shapes, but different colors than shown – models made before 2000 – are usually more common and comparatively less sought after, and therefore, worth less than the models in this book. There are exceptions, of course.

Secondly, value is greatly affected by condition. The Hartland 2000 horses were not marketed as toys, but as collector items. Most are likely to be – and remain for many years – in just about new condition. But, to be on the safe side, the values usually given are for very good, excellent, and near-mint condition. The condition abbreviations are: near mint = NM, excellent = EX, very good = VG, and average = AV. "Fair" and "poor" would be spelled out. Mint in package (still sealed in never-opened package) = MIP.

Given the small editions, the production models do not come up for sale very often. A general rule is that, if kept in new condition, your Hartland horses released since 2000 should be worth almost what you paid for them new. However, some colors and molds are more popular than others, and were able to sell

Hartland horses group nicely on a shelf. These production models from 2001 are the 9" Five-Gaited Saddlebred in pearled black with white points, 11" Arabian in dappled palomino, and 11" Quarter Horse in chocolate bay.

for more than their original price even within a few years after they were released. After enough time passes, even the less popular models will tend to climb in value given how few were made. So, I've factored that into the values so that most models are now "worth" a little more than their original price – but only if they're in nearly new condition, which would be about a condition score of about 92% or higher.

**Rating Condition.** A condition score of 60%-79% is very good; 80%-94% is excellent; and 95%-100% is near-mint to mint. (A score of 40%-59% would be average condition, but values for average condition are not included.) A score of 100% would be rare. Some models – and this is true of any brand of horse – already have very minor imperfections when they arrive: factory flaws or shipping damage. Then, things happen, even in a home with no children or free-roaming pets. Deduct points for damage as follows:

Missing an ear tip, subtract 10%, and the model's condition is excellent at best, depending on other damage that may be present.
Missing the top one-quarter of an ear, subtract 20%.
Missing the top one-half of an ear, subtract 25%.

With one-quarter to one-half of an ear missing, the model's condition is very good at best. Be sure to count damage on each ear separately. If both ear tips are missing, that is minus 20%, and the model's condition is no better than very good.

For rubs or scratches (paint or finish missing), factors to consider are location on the model's body, size, and depth: whether the rub or scratch affected only the top layer or went all the way down to the bare (usually, white) plastic. A related type of damage is marks added, such as when the horse fell against another horse or other object that "left its mark." Rubs or marks on the sides of the horse – what you see in a side view of the horse – are more unsightly than rubs or marks in peripheral areas, such as the ear tips, nose front, tail back, or bottom edges of the hooves. It's subjective. To simplify, visualize all of the rubs and marks grouped together, touching each other. How big an area would that occupy? Then, make deductions as follows.

If the rubs and marks add up to the area of:

A pencil point, subtract 6%, and the model's condition is excellent at best.
The bottom of a pencil eraser, subtract 35%, and the model's condition is very good at best.
The area of a dime, subtract 55%, and the model is average at best.

A horse with one or two pin-tip-sized rubs only on the periphery could be near-mint, but a single pin-head-sized rub in the side view would probably disqualify it from being near-mint, and move it down to excellent or lower, depending on other damage.

Being sold almost entirely to collectors (adults), the horses from Hartland 2000 got an excellent start in life, and most should still be in near mint condition – or at least very high in the excellent range.

# CHAPTER 1
# AMERICAN SADDLEBREDS
## 9" FIVE-GAITERS

*Five-Gaited Saddlebreds stepped past, keeping perfect time.* The parade of new 9" series Five-Gaited American Saddlebreds began in March 2001 with a pearled palomino. What a start! Hartland's 9" series Five-Gaiters are 8.25" tall, 9.5" long, and executing a slow gait: the stepping pace.

**As Pictured.** The eight production colors of 9" series Five-Gaiters, described in the order pictured, are:

### A

- Dapple Grey (Pearled) 9" Five-Gaiter, "Snow Angel," 881-06; December 2002, 62 made; VG: $21… EX: $28… NM: $35.

The 2002 Holiday Horse, "Snow Angel," is both dappled and pearled. The dapples are white in a field (body color) of light gray or silver. The four white socks appear silvery white because of the silver-pearl coat on much of the model. The black areas – muzzle, knees, and hocks – are not pearled, nor are the peach hooves, white star, and peach snip.

**A**
Dapple Grey (Pearled) 9" Five-Gaiter, "Snow Angel," 2002 Holiday Horse, 881-06.

**B**
Rose Grey (Dappled) 9" Five-Gaiter (Dapple Grey with Bay Shadings), 881-03.

**C**
Flaxen, Liver Chestnut 9" Five-Gaiter, 881-09.

**D**
Caramel Bay (Dappled) 9" Five-Gaiter, 881-05.

**E**
Palomino (Pearled) 9" Five-Gaiter, 881-01.

**F**
Black with White Points (Pearled) 9" Five-Gaiter, 881-02.

"Snow Angel" was announced in late November 2002, and could be ordered until January 31, 2003. The first ones were shipped shortly before Christmas. The original price was $30, and remained at $30 when most prices were reduced in January 2003. The factory's complete name for this model was: 2002 Holiday Horse "Snow Angel" Dapple Pearl Grey Heritage Series Saddlebred. (Maybe it's a coincidence, but Breyer's 2002 holiday model also was a grey Five-Gaiter [glossy white with gray shadings].)

"Snow Angel" was a numbered edition, and came with a decorative tag. A total of 62 were produced. Look for three digits in black ink on the model's underside; for example: #052.

## B

- Rose Grey (Dappled) 9" Five-Gaiter (Dapple Grey with Bay Shadings), 881-03; February 2002, 50 made; VG: $21... EX: $28... NM: $35.

This Saddlebred is dappled, but not pearled. It looks like a bay that has turned about 65% grey. Its body is white with extensive, pale brown shadings, and dapples in the shadings. The tail is tri-color: black on top and white below, with a pale brown transition. Otherwise, the points are black along with the hooves. Dark shading was freehand-airbrushed on the back. The face is white ("bald"), except for black shading on the muzzle. The factory called it: dapple rose grey.

This model debuted in February 2002 as an edition of 50. The original price was $36, reduced to $29 in January 2003. (From November 15-December 20, 2002, the price was $25.50.) In August 2004, three remained; they were sold by mid-November 2004.

## C & L

- Flaxen, Liver Chestnut 9" Five-Gaiter, 881-09; May 2006, 25 made; VG: $27... EX: $36... NM: $45.

This model has a golden beige (flaxen) mane and tail and medium brown (liver chestnut) body; three white stockings above peach hooves; and a black right front hoof. A painted-on white blaze ends in peach color between the nostrils. The white stockings are unpainted plastic. The model is not pearled.

A run of 25 in May 2006, nine were still available in March 2007. The original price was $30. The factory called the color: liver chestnut.

## D

- Caramel Bay (Dappled) 9" Five-Gaiter, 881-05; July 2002, 50 made; VG: $27... EX: $36... NM: $45.

This model's body has very pale, golden-brown dapples within light brown shading, and appears, overall, as a subdued caramel color. The mane, tail, and right foreleg and hoof are black. The other three legs have white stockings and peach hooves below black knees and hocks. The stockings have a masked top edge. A white blaze ends in peach color like the hooves. The company website mentioned "a slight pearl tint."

The factory originally called the color: dapple bay. In my newsletters, I called it "dapple bay (caramel color)" or "bay with dapples (caramel color)." The company followed suit, eventually listing it as: dapple caramel bay.

This bay, an edition of 50, was announced in late June 2002, and arrived in July 2002. The original price was $34, reduced to $29 in January 2003. (From November 15-December 20, 2002, it was on sale for $25.50.) Eight were left in December 2003, and one was left in November 2005.

## E

- Palomino (Pearled) 9" Five-Gaiter, 881-01; March 2001, 150 made; VG: $27... EX: $36... NM: $45.

One of the very first models from this company, the palomino ("golden palomino") debuted in March 2001. The edition size announced in advance was 500, but was changed to "unlimited" when the model came out, and reduced to 250 by August 2001. It was finally capped at 150. The original price was $36, reduced to $30 in January 2003. (From November 15-December 20, 2002, it was $27.)

This model seems to blend all the best colors one could hope to see in a palomino model. Its golden-yellow body has pale, reddish-gold shadings, and is subtly reflective (pearled). The white mane and tail are obviously pearled. The four white stockings are pearled from about the fetlocks up while the pasterns are plain white (not pearled). The stockings do not have a distinct, upper edge; instead, they emerge from beneath the golden color, which was airbrushed over the top. The hooves are peach color.

In August 2004, two pieces remained; they were sold by mid-November 2004. (For two more photos, see the Introduction.)

# F

- Black w/ White Points (Pearled) 9" Five-Gaiter, 881-02; October 2001, 50 made; VG: $24... EX: $32... NM: $40.

This pearled Five-Gaiter was an edition of 50 announced in September 2001. It arrived in October 2001, and was sold out by mid-January 2002. The price was always $36.

A fantasy color, the company called it: silver charcoal. The black body color has tiny silver specks that catch the light. This pearled effect is also obvious on the white mane and tail, and present, but less apparent, on the four white stockings and peach hooves. The horse has a masked-off white blaze painted on. The blaze, which also looks pearled, turns to pink between the nostrils.

**G**
Chestnut Pinto 9" Five-Gaiter, "Wave the Banner," 2002 Club Model, 881-04.

**H**
The Chestnut Pinto 9" Five-Gaiter, "Wave the Banner," has a reddish brown tail tip and black feet.

**I**
Bay Pinto 9" Five-Gaiter, 881-07.

**J**
The Bay Pinto 9" Five-Gaiter (881-07) has beige-pink feet, darker than the usual Hartland 2000 hoof shade.

**K**
Six of the 9" Five-Gaiters have face markings: the Black (881-02), Chestnut Pinto (881-04), Caramel Bay (881-05), Dapple Grey, "Snow Angel" (881-06), Bay Pinto (881-07), and Liver Chestnut (881-09).

**L**
The 9" Five-Gaiter in flaxen, liver chestnut (881-09) has three white stockings and one black hoof.

## G & H

- Chestnut Pinto 9" Five-Gaiter, "Wave the Banner," 881-04; January 2002, 130 made; VG: $21... EX: $28... NM: $35.

"Wave the Banner" was the club model in the $50 (ppd.) membership package for 2002. A lapel pin with the model's picture on it was also included.

About 65% white and 35% reddish chestnut, this model's pattern of tobiano pinto markings was created by adding the brown (burnt orange) paint. The white areas, which have a shine, are the unpainted, white plastic. An exception is the large star on the horse's forehead; there, white paint was applied over brown paint. All marking edges were masked. The hooves are black. The model is not pearled.

The factory called the color: chestnut pinto. A total of 130 were produced. "Wave the Banner" was a numbered edition marked in black ink on the model's underside; for example: 062. The model was pictured on the 2002 Hartland Horses Collector's Club lapel pin.

## I & J

- Bay Pinto 9" Five-Gaiter, 881-07; December 2004, 104 made; VG: $18... EX: $24... NM: $30.

This pinto is more brown than white. It appears that the white, tobiano pinto markings were painted on, rather than merely being the bare plastic. White paint was hand-brushed on the legs, but airbrushed (over brown paint) on the body. This model was not pearled.

When it was released in December 2004, the company said there were fewer than 50, but the final tally was 104. In January 2007, 47 were left, according to the company newsletter.

The factory called the bay pinto a "sample run" – the work of a different painting service than before – and priced the model at $19.99 due to slight painting flaws. Details include a peach spot on the lower lip and a white blaze. The hooves are a darker shade of peach than the other Five-Gaiters. All four hoof bottoms were painted.

| AMERICAN SADDLEBREDS: 9" SERIES FIVE-GAITERS – PRODUCTION MODELS | | | | |
|---|---|---|---|---|
| **Color & Name** | **Item #** | **Run Type** | **Debut Date** | **Qty.** |
| Palomino (Pearled) | 881-01 | r | March 2001 | 150 |
| Black with White Points (Pearled) | 881-02 | r | October 2001 | 50 |
| Chestnut Pinto, "Wave the Banner" | 881-04 | Club model, 2002 | January 2002 | 130 |
| Rose Grey (Dapple Grey with Bay Shadings) | 881-03 | r | February 2002 | 50 |
| Caramel Bay (Dappled) | 881-05 | r | July 2002 | 50 |
| Dapple Grey (Pearled), "Snow Angel" | 881-06 | Holiday horse, 2002 | November 2002 | 62 |
| Bay Pinto | 881-07 | r | December 2004 | 104 |
| Flaxen, Liver Chestnut | 881-09 | r | May 2006 | 25 |

**9" Five-Gaiters** – The Five-Gaiters spanned six years (2001-2006), and included the 2002 club model and 2002 holiday horse, and six regular runs, which are denoted by "r" under "Run Type". Of the eight production colors, three were noticeably pearled. Three editions were dappled, and two were pintos. Three editions were larger than 100.

**Saddlebreds with Style.** In 2001, Hartland's graceful, 9" series Five-Gaiters returned to the spotlight in style. Pearled colors led the way, but realism worked in tandem with fantasy. The finale, in 2006, was a textbook flaxen chestnut.

# Chapter 2
# Rearing Mustangs
## 9" Woodcuts

*Mustang stallions reared and wheeled in mock fights over imaginary mares.* This mold looks whittled from wood (but wasn't), and is called the woodcut Mustang (or Mustang woodcut) to distinguish it from the smooth-surfaced rearing Mustangs, which were last made in the 1990s. Part of the 9" series, the rearing woodcut Mustang measures 9.5" tall and 8.75" long.

**Six Production Colors**. The captions label each Mustang as 9" and woodcut, but those terms are omitted from the descriptions that follow because all Hartland Mustangs ever made are 9" scale, and all of the Mustangs made since 2000 are woodcuts. The six colors of production Mustangs since 2000, described in the order pictured, are:

**A**
Buckskin Pinto 9" Mustang woodcut (Pearled, with Black Tail), 886-03B.

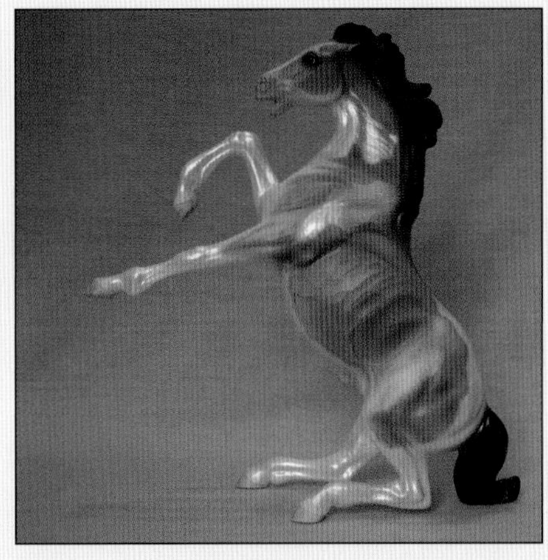

**B**
Buckskin Pinto 9" Mustang woodcut (Pearled, with Black Tail), 886-03B) – *left side.*

**C**
The Buckskin Pinto 9" Mustang woodcut (Pearled, with Black Tail, 886-03B) has a partial dorsal stripe.

**D**
Blue Roan 9" Mustang woodcut, 886-06.

## A, B, & C

- Buckskin Pinto Mustang w/ Black Tail (and Pearled), 886-03B; August 2001, 21 with black tail; VG: $30... EX: $40... NM: $50.

For Jamboree 2001 (June 22-23, 2001), 50 Mustangs were painted buckskin pinto with a white tail (and black mane). Some were given to Jamboree volunteers. Then, the company painted the tail black on the rest, and in August 2001, sold them for $32 to winners of a mail-in drawing through the company newsletter. Subscribers entered by mailing in their name and address.

Two company newsletters at the time reported 29 white-tailed gift models and 21 black-tailed sold (production) models, but the company later published the opposite: 21 white and 29 black. I lean toward the earlier reports being correct.

This Mustang has a pale, golden-tan body with peach feet and muzzle, and white tobiano pinto markings. The mane and tail are entirely black. The golden areas are matte finish. The white areas are pearled white. Collector Tammy Nguyen described the eyes as black with a white back corner and a white dot highlight in the black area. *Photos A & B, courtesy of Marla Phillips. Photo C: model, courtesy of Tammy Nguyen; photo, courtesy of Robyn Porter.*

## D & N

- Blue Roan Mustang, 886-06; October 2003, 25 made; VG: $21... EX: $28... NM: $35.

The Blue Roan Mustangs look like they began with a coat of black paint, and then white paint was sprayed over the body, neck, and upper legs. The points are black, and the body color simulates the mixed black and white hairs on a blue roan horse. The finish is not pearled. The model's face has a large, masked star with a pointed bottom.

The Blue Roan Mustang was a surprise, special run at the 2003 Jamboree (October 24-26, 2003). Of 25 painted, 22 were available for $30 at Jamboree and sold out. The other three were prizes mailed after Jamboree to the company's Hoof Pick Contest winners announced in its Summer 2003 newsletter (which arrived in November 2003). *Photo N, face picture, courtesy of Melanie Teller.*

## E, F, & G

- Palomino Pinto Mustang (Pearled), 886-05; December 2001, 50 made; VG: $33... EX: $44... NM: $55.

This model has a pale, golden beige-yellow body with white, tobiano pinto markings; the muzzle and hooves are peach. Its finish is pearled from head to foot; everything, except the eyes, is pearled! This Mustang debuted in December 2001 as an edition of 50. Some were still available on January 12, 2002, but not for long. The original price was $34.

**E**
Palomino Pinto (Pearled) 9" Mustang woodcut, 886-05.

**F**
Palomino Pinto (Pearled) 9" Mustang woodcut, 886-05 – *left side.*

**G**
The Black-Gray Pinto and Palomino Pinto 9" Mustangs have the same pattern of white markings, and so does the Buckskin Pinto (photo C).

**H**
Black (Pearled) 9" Mustang woodcut, 886-04. The points and hooves are regular black; the body is pearled black.

**I**
The Black-Gray Pinto (Pearled) 9" Mustang woodcut, 886-02, June 2001, varied from near-black *(left)* to charcoal gray.

**J**
These are darker and lighter versions of the Black-Gray Pinto Mustangs (886-02) – *left side*. This example of the darker one does not have the left hip spot.

**K**
Chocolate Bay 9" Mustang woodcut, 886-01.

**L**
This example of the darker version of the Black-Gray Pinto Mustang (886-02) has a white spot on its left hip.

## G, I, J, & L

- Black-Gray Pinto (Pearled) Mustang, 886-02; June 2001, 50 made: VG: $27... EX: $36... NM: $45.

The company called this color, "silver black pinto" or "grey pinto." It debuted at Jamboree, June 22-23, 2001, moved to the company website, and was sold out by mid-July 2001. The price was $32.

The Black-Gray Pinto Mustang is found in lighter and darker variations: one medium gray and one dark gray, dark enough to be mistaken for black. Collector Marla Phillips described hers: "Both of them are pearly in the white areas." On her lighter version, the gray areas are pearly, but not as much as the white areas, she said. "It sort of has a semi-gloss look to it like the pearl coat was added on top of the gray. The darker one is more of a flat matte gray, but it is still slightly pearled when the light hits it right." Marla's darker version is also missing the white spot from its left hip, and has no eye whites.

On my dark version, the pearling ends on the pasterns, and the lower pasterns are plain white (not pearled). The peach hooves and muzzle are also not pearled. The white areas are reflective and obviously pearled, evidently with tiny silver flecks in the pearled coat.

Before 2000, the Hartland Mustangs in black pinto and grayish pinto always had black hooves and non-pearled finishes. The post-2000 Mustang pintos, with their pink feet and pearled coats, are a breed apart. *Photos I & J, courtesy of Marla Phillips. Photo L, courtesy of Eleanor Harvey.*

## H

- Black (Pearled) Mustang, 886-04; January 2002; 11 made; VG: $36... EX: $48... NM: $60.

The black Mustang with pearled finish was sold through a drawing in the company newsletter. Subscribers had about a month to mail their contact information on a 3" x 5" card. The deadline was January 15, 2002. The model was shipped to 11 winners who each paid $34. It is the rarest model that is counted as a production run.

The body is pearled black, but the points (mane, tail, and lower legs) and the hooves are regular black, according to collectors Robyn Porter, Tammy Nguyen, Elaine Boardway, and April Powell, who each own one. The black pearl Mustang's eyes are black with a white back corner and white dot, Robin said. Its face is black with no white markings. *Photo H, courtesy of April Powell.*

## K

- Chocolate Bay Mustang, 886-01; June 2001, 100 made; VG: $21... EX: $28... NM: $35.

The Chocolate Bay Mustang debuted at Jamboree in June 2001 as a surprise special run of 50 for $32 each. The edition size increased to 100 by October 2001, and in January 2003, the price was reduced to $27. (From November 15-December 20, 2002, the price was briefly $24.) In November 2004, there were seven left, and by November 2005, they were sold out.

This model has a medium-to-dark brown body with black points, hooves, and muzzle; a large, narrow star and small stripe on the bridge of its nose; and a semi-gloss finish. It is molded in opaque, snowy white plastic, visible on the bottom of the hind feet. The finish is not pearled. The company always called it: chocolate bay. *Photo K, courtesy of Eleanor Harvey.*

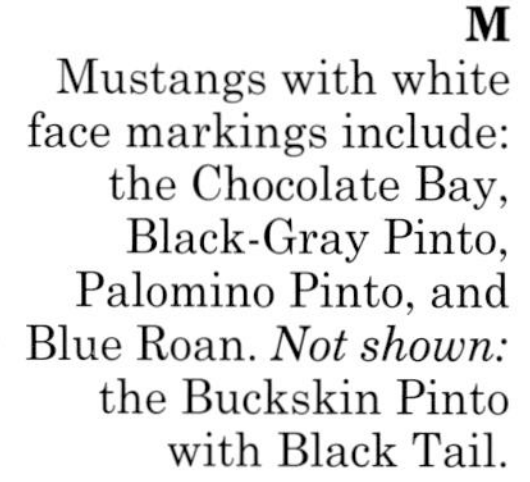

**M**
Mustangs with white face markings include: the Chocolate Bay, Black-Gray Pinto, Palomino Pinto, and Blue Roan. *Not shown:* the Buckskin Pinto with Black Tail.

**N**
The Blue Roan Mustang has a large (white) star on its face, and pale peach "eye whites" painted into the front corner.

| Rearing Mustangs: 9" Series Woodcuts - Production Models | | | | |
|---|---|---|---|---|
| **Color** | **Item #** | **Run Type** | **Debut Date** | **Qty.** |
| Chocolate Bay | 886-01 | r | June 2001 | 100 |
| Black-Gray Pinto (Pearled) | 886-02 | r | June 2001 | 50 |
| Buckskin Pinto with Black Tail | 886-03B | Sold to winners of mail-in drawing | August 2001 | 21 |
| Palomino Pinto (Pearled) | 886-05 | r | December 2001 | 50 |
| Black (Pearled) | 886-04 | Sold to winners of mail-in drawing | January 2002 | 11 |
| Blue Roan | 886-06 | Jamboree SR of 22 (+ 3 as Gifts) | October 2003 | 25 |

**Mustangs** – Of the six production colors of Mustangs, three were regular runs (denoted by an "r" under "Run Type"), and three were special runs not available in the usual ways. All were released in 2001-2003. The largest edition, at 100, was the Chocolate Bay.

**Hard to Get.** The Mustangs turned loose from 2001 through 2003 were a challenge to capture. The blue roan sold out at a weekend event. The black-gray pinto and palomino pinto were gone in two months. The pearled black and the buckskin pinto with black tail required the luck of the draw. Only the chocolate bay was loose on the prairie for years, waiting to be won over with carrots and a kind word.

# Chapter 3
# 9" Tennessee Walkers
## (Woodcuts)

*Tennessee Walkers nodded along.* The 9" series Tennessee Walkers are woodcuts with a whittled-like surface. There were two mold variations, and the ones produced since 2000 are the type with mane hanging down on both sides of the neck. They are 8.25" tall, 9.5" long, and glide along at a running walk.

**Seven Colors.** The 9" Tennessee Walker was painted in seven production colors. They are described in the order pictured. Two of them had names.

**A**
Copper Chestnut with White Socks 9" Tennessee Walker, 887-04.

**B**
Black with White Socks 9" Tennessee Walker, 887-02.

**C**
Buckskin 9" Tennessee Walker, 887-06.

**D**
Bay 9" Tennessee Walker, "Lion Hart," 2006 Club Model, 887-07

**E**
Dapple Grey with White Socks (Pearled) 9" Tennessee Walker, "Noel," 2001 eBay Holiday Horse, 887-01.

**F**
Dapple Grey with Black Points (Pearled) 9" Tennessee Walker, 887-05.

## A

- Copper Chestnut 9" Tennessee Walker, 887-04; July 2002, 50 made; VG: $24... EX: $32... NM: $40.

This model was an edition of 50 announced in late June 2002; it arrived in July 2002. Priced at $36, it was reduced to $30 in January 2003. (From November 15-December 20, 2002, it was briefly $27.) Four were left in November 2006, and in March 2007, two were left.

The company called it: copper chestnut. Its body paint has a very matte texture. The color is similar to copper color, but lighter. It is not a "metallic copper" color, but it does have a slight glow, perhaps because a pearled white coat lies beneath. The four white stockings are pearled (reflective); the hooves are peach. The mane, knees, hocks, and upper part of the tail are a darker brown than the body. The model's masked-off white blaze (plain white, not pearled) ends in pink at the nostrils.

## B

- Black w/ White Socks 9" Tennessee Walker, 887-02; December 2001, 50 made; VG: $24... EX: $32... NM: $40.

The black color on this Walker appears plain black, but the four white socks are pearled. The masked, white star painted on the forehead is plain white (not pearled). The company called the color: black or raven black.

This model, an edition of 50, came out in December 2001 for $36, reduced to $30 in January 2003. (From November 15-December 20 2002, it was temporarily $27.) Six were left in December 2003. They sold out between December 2005 and May 2006.

## C

- Buckskin 9" Tennessee Walker, 887-06; December 2004, 58 made; VG: $21... EX: $28... NM: $35.

This model's body is an even and glossy, light golden-tan buckskin color with glossy, black color on his mane, tail, knees, hocks, groin, muzzle, and lower legs except for low (white) socks on all four legs. The color is not pearled. He was called a "sample run" because the company tried a different painting service. The hooves are a darker shade of peach than usual.

He went up for sale in December 2004 for $19.99. A total of 58 were made. In March of 2007, 13 were left. He was sold out between November 2007 and October 2008.

## D & I

- Bay 9" Tennessee Walker, "Lion Hart," 887-07; January 2006, 50 made; VG: $27... EX: $36... NM: $45.

"Lion Hart" was the 2006 club model, breed series option. (The $55 membership also included a key chain made from a Hartland Dale Evans hat.) This bay stallion was limited to 50 pieces. Early in 2007, two pieces were left, but they didn't last long.

This model's body appears as a blend of warm browns, probably less because different colors were used, than because the airbrush was trained longer on some areas than on others. The color is darker over the back. "Lion Hart" has three low, white socks; otherwise, his lower legs are black along with the knees, hocks, mane, and tail. Three hooves are peach, and one (the left fore) is black. His face has a star and pink snip. His finish is not pearled. He's a numbered edition; his underside has: #, followed by two digits.

## E

- Dapple Grey w/ White Socks (Pearled) 9" Tennessee Walker, "Noel," 887-01; November 2001, 85 made; VG: $24... EX: $32... NM: $40.

"Noel," the first 9" Tennessee Walker from Hartland 2000, is pearled, silvery gray with small dapples and four silvery-white stockings. The mane and upper half of the tail are medium-to-dark gray while the lower half of the tail is silvery white. The hooves are dark gray. (Contrast him with the model in photo F, which has black points.)

"Noel" was the 2001 eBay Holiday horse. He was $45 plus $6.95 shipping. The company called his color: dapple grey pearl. Each "Noel" order was accompanied by one pale gold Tinymite, "Glisten."

All six Tinymite breeds were painted pale gold. There was no choice of breed.

In mid-November 2001, the company said the edition would be limited to 250, but it was a tough holiday season for retail sales, and 14 months later, in January 2003, "Noel" was still available, then for $38. When he sold out between August and mid-November of 2004, the final run size was 85.

"Noel" is a numbered edition with a three-digit sequence number written in silver ink on his underside; for example: 033. He came with a decorative hang tag, too.

The first piece from the "Noel" run, numbered "#001" on the belly, was sold on eBay, December 7, 2001, with all six breeds of "Glisten" included. That was the only time anyone could acquire all six of the pale gold Tinymites. The winning bid was $260. (For more on "Glisten," see the Tinymite chapter.)

## F

- Dapple Grey w/ Black Points (Pearled) 9" Tennessee Walker, 887-05; October 2004, 25 made; VG: $24... EX: $32... NM: $40.

This Walker has white dapples on a light, silvery gray body, and black points and hooves. He is silver-pearled, but not on the black areas. The factory might have made him by adding black points to "Noel" models. (On mine, the dapples are less obvious than Noel's, but that might reflect variation within the two model runs, rather than a consistent difference between the two runs.)

This model was a surprise, special run of 25 at the 2004 Jamboree (October 8-10, 2004) and sold out there. He was $30 plus California sales tax.

## G

- Blue Roan 9" Tennessee Walker, 887-03; June 2002, 50 made; VG: $24... EX: $32... NM: $40.

To simulate blue roan color, it appears that light blue paint was lightly sprayed over a black undercoat. The mane, tail, and head are black, and black on the legs extends nearly up to the body line. All four hooves are black although the right foreleg and left hind leg have white socks. The model has an apron face, and its right eye is blue because it's within the white area. The model on the company's catalog sheet looks lighter, more like blue-gray, and less like blue-black, than the one in photo G.

The Blue Roan 9" Tennessee Walker, an edition of 50, debuted in June 2002. The price was $36, reduced to $30 in January 2003. (From November 15-December 20, 2002, it was briefly $27.) Two were left in December 2003, and they were gone by May 2004.

**G**
Blue Roan 9" Tennessee Walker, 887-03. Some may be lighter in color than this example.

**H**
The 9" Tennessee Walkers with face markings are: the Black (887-02), Blue Roan (887-03), Copper Chestnut (887-04), Buckskin (887-06), and Bay (887-07).

**I**
Bay 9" Tennessee Walker ("Lion Hart," 887-07) – *right side.* The mold is called a woodcut because it looks whittled.

| TENNESSEE WALKERS: 9" SERIES WOODCUTS – PRODUCTION MODELS | | | | |
|---|---|---|---|---|
| **Color & Name** | **Item #** | **Run Type** | **Debut Date** | **Qty.** |
| Dapple Grey with White Socks (Pearled), "Noel" | 887-01 | eBay Holiday horse, 2001 | November 2001 | 85 |
| Black with White Socks | 887-02 | r | December 2001 | 50 |
| Blue Roan | 887-03 | r | June 2002 | 50 |
| Copper Chestnut with White Socks | 887-04 | r | July 2002 | 50 |
| Dapple Grey with Black Points (Pearled) | 887-05 | Jamboree 2004 surprise special | October 2004 | 25 |
| Buckskin | 887-06 | r | December 2004 | 58 |
| Bay, "Lion Hart" | 887-07 | Club model, 2006 | January 2006 | 50 |

**9" Tennessee Walkers** – Seven colors of 9" Tennessee Walkers waltzed through the production line from 2001-2006. They included: four regular runs, a holiday horse, a club model, and a surprise special run. The largest edition, at 85, was "Noel," a 2001 holiday horse.

Hartland 2000 was the first Hartland company to issue the 9" woodcut Tennessee Walkers in "real horse" and/or detailed colors. In the 1960s and 1970s, they were only finished in simulated wood colors – dark brown, black, and tan, and did not have details like the eyes painted in.

# Chapter 4
# 7" Tennessee Walkers
## (Family Series)

*As handlers led them, mares kept an eye on their foals.* In the 7" series Tennessee Walker Family, the mare and stallion glide along at a pleasure-type running walk while the foal stands with its front legs braced and left hind hoof on tip. The stallion is 6.5" tall x 8" long; mare, 6" x 7.5," and foal, 5" x 4.25." They are smooth-textured (not woodcuts).

**A**
Black Pinto 7" Tennessee Walker Stallion, 684S-01.

**B**
Black Pinto 7" Tennessee Walker Stallion – *left side.*

**C**
7" Tennessee Walker Foals in Black Pinto, 684F-03, and Bay Pinto, 684F-02.

**D**
7" Tennessee Walker Foals in Black Pinto and Bay Pinto – *left side.*

**Sold Individually.** Two stallions and two foals were sold individually, in colors different from the family sets. All four have a regular, not pearled, finish.

## A & B

- Black Pinto 7" Tennessee Walker Stallion, 684S-01; April 2001, 125 made; VG: $15... EX: $20... NM: $25.

This stallion is a black-and-white, tobiano pinto with a high, white sock on his left foreleg and low sock on his left hind leg. The hooves under the socks are peach; the other two feet are black.

His edition size changed often. It was 500 when he was announced in March 2001, 250 by August 2001, and 100 by May 2004. Then, it rose to 150 by August 2004, and was finally capped at 125 by January 2007. The original price was $25, reduced to $20 in January 2003. (It was briefly $18.75 from November 15-December 20, 2002.) In February 2009, 19 were left. Some remained in December 2010, but were gone in the first months of 2011. *Photos A & B, courtesy of Eleanor Harvey.*

## C & D (left Model)

- Black Pinto 7" TW Foal, 684F-03; February 2002, 50 made; VG: $11... EX: $14... NM: $18.

The black-and-white, tobiano pinto 7" TW Foal has four white stockings and peach feet, and looks related to the stallion. (Her color, like his, is plain black, not blue-black or pearled black. No production-color mare matches either the black

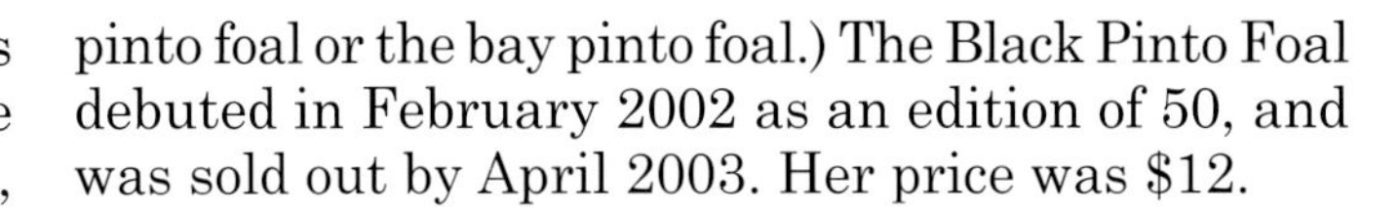

pinto foal or the bay pinto foal.) The Black Pinto Foal debuted in February 2002 as an edition of 50, and was sold out by April 2003. Her price was $12.

## C & D (Right Model)

- Bay Pinto 7" TW Foal, 684F-02; February 2002, 50 made; VG: $11... EX: $14... NM: $18.

The bay pinto 7" TW Foal looks like the black pinto foal except that she's a subdued, light brown with dark brown knees, hocks, and mane. Her tail is almost black; her points may be black after her foal coat sheds out. She was $12, and sold out by April 2003.

## E

- Dapple Grey 7" TW Stallion, 684S-03; September 2002, 50 made; VG: $15... EX: $20... NM: $25.

This stallion has fine, white dapples in gray shading over a white body; black hooves and lower legs; and dark gray paint on the upper tail and on the mane, but only along the crest of the neck. He debuted in early September 2002 as an edition of 50. His price, $25, was reduced to $20 in January 2003. (From November 15-December 20, 2002, he was briefly $18.75.) In January 2007, seven were left. They were sold out by February 2009.

**7" Tennessee Walker Family Sets.** Since 2000, Hartland made three-piece Tennessee Walker family sets in six colors. Those families are:

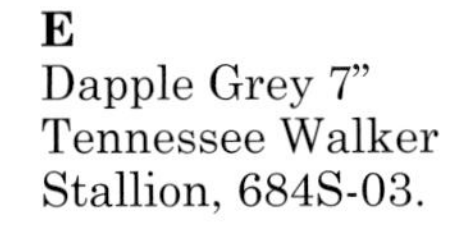

**E**
Dapple Grey 7" Tennessee Walker Stallion, 684S-03.

**F**
The 7" and 9" Tennessee Walker stallions have similar poses, but the smaller stallion's chin is more raised, and his tail flows backward more.

## G & N

- Champagne (Pearled) 7" TW Family, 684-01; March 2001, 150 made; VG: $37... EX: $50... NM: $62.

Pearled color enhances this set, which was among the first horses from Hartland after 2000. The company called this family "amber champagne," but calling it just-plain "champagne" is a better fit because there are several varieties of champagne, including amber, classic, and gold, and different members of this family approximate different types. (Read more later in this chapter.)

The Champagne Tennessee Walker Family began in March 2001 as an unlimited edition, and then was downsized: to 250 by October 2001, to 200 by August 2004, and finally, to 150 by May 2006. The original price was $45, reduced to $38 in January 2003. (From November 15-December 20, 2002, it was briefly $33.75.) Twelve sets remained in January 2007. By February 2009, seven sets were left, and the price was $28. By December 2010, the price was $25. Some additional sets must have turned up in the warehouse, because in July 2011, more than 10 were available. That was still true in January 2012. *Photo N, courtesy of Marla Phillips.*

Individually, values are: Stallion – VG: $14... EX: $18... NM: $22; Mare – VG: $14... EX: $19... NM: $23; Foal – VG: $9... EX: $13... NM: $17.

## H

- Buckskin 7" TW Family, 684-03; January 2002, 50 made; VG: $33... EX: $44... NM: $55.

The Buckskin TW Family was an edition of 50 in February 2002, priced at $45. The price went down to $38 in January 2003. (From November 15-December 20, 2002, it was $33.75.) Seven sets remained in November 2006; two were left in February 2009 for $28 each.

**G**
Champagne (Pearled) 7" Tennessee Walker Family, 684-01. The foal's body is pale taupe; the stallion's is pale, coppery taupe; the mare's is tan.

**H**
Buckskin 7" Tennessee Walker Family, 684-03. The stallion *(right)* lifts his knee higher than the mare. The parents have black points; the foal has brown points.

The set is pale beige-yellow with a smooth finish. Their color is neither dappled nor pearled.

The mare has black points and hooves. The stallion has black points and hooves except for a left hind sock over a peach hoof. Both the mare and stallion have black on their muzzle. The foal has black nostrils; brown mane, tail, knees, and hocks; four white stockings; peach hooves in front, and dark brown hooves in back.

Their individual values are: Stallion – VG: $13... EX: $16... NM: $20; Mare – VG: $12... EX: $16...$NM: $20; Foal – VG: $8... EX: $12... NM: $15.

The Black TW Family is both dappled and pearled. The company called it: dapple black pearl. The dapples are numerous, small, and silvery-colored in a field of shiny, pearled black. The stallion has a short (white) blaze on its face; the mare has a short blaze that is longer and shaped differently than the stallion's blaze.

This family debuted in July 2001 as an edition of 50. Its price was $45. It sold out by April 2002.

Values for each family member are: Stallion – VG: $14... EX: $19... NM: $23; Mare – VG: $14... EX: $18... NM: $22; Foal – VG: $9... EX: $13... NM: $17.

## I

- Black (Dappled and Pearled) 7" TW Family, 684-02; July 2001, 50 made; VG: $37... EX: $50... NM: $62.

## J & M

- Bay (w/ Pearled Undercoat) 7" TW Family, 684-04; August 2003, 25 made; VG: $37... EX: $50... NM: $62.

**I**
Black (Dappled and Pearled) 7" Tennessee Walker Family, 684-02.
All four of the mare's feet (she's *at left*) touch the ground.

**J**
Bay (Sub-pearled) 7" Tennessee Walker Family, 684-04. The foal has pearled white stockings.
In this example, the foal is paler than his parents, and the mare is redder than the stallion.

**K**
Bay Roan 7" Tennessee Walker Family, 684-06. The mare *(left)* seems to be watching her foal; her head is carried lower than the stallion's.

The Bay TW Family debuted in August 2003 for $38. The edition size was 25. In February 2009, two sets remained, and the price was $28.

The company called them "bay," "chocolate bay," or "shaded bay." The medium, warm brown, bay body color appears to be painted over a pearl white coat, which adds an under glow. The black areas on the models are not pearled.

The mare and stallion have black points (mane, tail, and lower legs) and hooves. The foal's mane, tail, knees, hocks, and hooves are black; her four white stockings are pearled, and the pearled coat shows more on her body than on theirs. The foal also has a small, round (white) star neatly painted, perhaps freehand, over the brown paint on its forehead. My mare is a slightly redder shade of bay than the stallion and foal, which look browner. Other sets might reflect slight color variation also. *Photo M, courtesy of Marla Phillips.*

Individually, values are: Stallion – VG: $14... EX: $18... NM: $22; Mare – VG: $14... EX: $19... NM: $23; Foal – VG: $9... EX: $13... NM: $17.

## K

- Bay Roan 7" TW Family, 684-06; June 2006, 25 made; VG: $30... EX: $40... NM: $50.

The Bay Roan family members have black points and a brown head. Their slightly rosy, near-white body represents the mix of reddish brown and white hairs on a bay roan. The head and points of a roan remain the un-roaned color, so that's why the head is brown, the points are black, and there's brown above the knees and hocks, where the black points leave off. The mare has a (white) blaze ending above the nostrils (or a wide and long, joined star and stripe).

The company called this color: bay roan. It was not pearled. The Bay Roan TW Family was a run of 25 in June 2006. The price was $38. In February 2009, two sets were still available, and the price was $28. They were sold out by December 2010.

Individually, suggested values are: Stallion – VG: $11... EX: $15... NM: $18; Mare – VG: 11... EX: $14... NM: $18; Foal – VG: $8... EX: $11... NM: $14.

## L

- Chestnut Roan 7" TW Family, 684-05; June 2006, 25 made; VG: $30... EX: $40... NM: $50.

The Chestnut Roan family has rust-colored points, and a rosy body to represent the roan mix of chestnut and white hairs. The mare and foal have a rust-colored head, while the stallion's head is more brown than rust. The stallion has a short blaze like the one on the black, 7" series TW family stallion. The stallion has a left front stocking with a peach hoof below it. All other hooves on the set are black (or dark).

The company called this set: strawberry roan. It produced 25 sets in June 2006. The original price was $38. In February 2009, there were nine sets still available for $28 each. They were sold out by December 2010.

Individual values are: Stallion – VG: $11... EX: $15... NM: $18; Mare – VG: 11... EX: $14... NM: $18; Foal – VG: $8... EX: $11... NM: $14.

**L**
Chestnut Roan 7" Tennessee Walker Family, 684-05. The Chestnut Roan family has rust-colored points while the Bay Roan family has black points.

## More on the Champagne 7" Tennessee Walker Family

Here's more description of the Champagne Family Tennessee Walkers in photo G.

**Stallion.** The stallion's body is pale, coppery taupe, lighter than the foal's. His mane and tail are copper brown, with darker color (maybe gray) over the upper tail and much of the mane. He has copper shading on his knees, hocks, and back; and four, pearled white stockings.

As he appears in photo G, his body and point colors closely match the photo of a horse "typical of the lighter end of classic champagne" on page 162 in *Equine Color Genetics, Third Edition* by D. Phillip Sponenberg, D.V.M., Ph.D. (2009).

Classic champagne body color is typically more like pale brown than the pale tan or yellow body color of amber champagne. Both varieties have brown points that are darker than the body (except that, on some classic champagne horses, the points match the body). In person, the stallion's body color looks warmer than in photo G, and more like amber champagne. So, he approaches both light classic champagne and dark amber champagne.

**Foal.** Champagne foals are typically darker as foals than they will be as adults, which is the opposite of what's true for most other horse colors. So, the baby in this family could lighten!

Right now, the foal has an iridescent taupe body, mane, and tail, with slightly darker brown knees and hocks, and silvery-white lower legs (white stockings). Her color is darker than the stallion's. Of the champagne colors, she comes closest to classic champagne.

**Mare.** The mare has warm brown over the top of a pearl white layer. Her mane and tail and lower legs are mainly pearled white. She comes close to being gold champagne, but the coat would be less reddish.

**Details.** All three models have dark feet that look painted brown and then painted black over the brown. All or most areas of the foal and stallion are pearled (over the top); on the mare; the brown layer was sprayed over the pearled layer. On all three, the visible white areas are luminous, pearled white, with non-distinct (unmasked) boundaries. The reflective, pearled finish can make it hard to capture the colors in a photograph.

Despite the attractiveness and the detail, there is a technical problem. The mare and stallion have a black muzzle (and the foal's muzzle is gray). That means they have dark skin. (The black or dark shading in the groin confirms that.) All three of them have black eyes. Champagne horses, though, have light eyes and pink skin.

**M**
In this example of the Bay (Sub-pearled) 7" Tennessee Walker Family (684-04), the foal matches the stallion, and the mare is a little redder shade of bay than they are.

**N**
It could be the photography, but in this Pearled Champagne 7" Tennessee Walker set, the foal looks darker than some, and the mare looks lighter (less reddish).

**O**
Among the 7" family-series Tennessee Walkers, six have white face markings, *from left*: the Chestnut Roan Stallion, Black Stallion, Black Mare, Bay Roan Mare, Black Foal, and Bay Foal.

**Champagne Horse Color.** The skin color on most horses is dark gray or brown, with pink skin under white markings. Champagne horses, though, have pink skin everywhere although it often becomes freckled with tan spots that might merge somewhat. Most horses have dark brown eyes, but champagnes typically have amber or hazel eyes (but blue in foals.)

The champagne gene, a simple dominant, lightens red pigment to gold and black pigment to brown. So, with no other modifiers present, a bay horse becomes golden with brown points ("amber champagne"); a black horse becomes taupe to chocolate brown with points matching, darker, or mixed ("classic champagne"); and a chestnut or sorrel becomes golden with a white, lighter, or matching mane and tail ("gold champagne").

Oddly enough, a gold champagne horse can have a white mane and tail, but knees and hocks that are a little darker than the body.

Some of the horses that used to be identified as buckskins, duns, and palominos; that is, the ones with pinkish skin, are now classified as champagne.

Sources: *Equine Color Genetics, Third Edition*, by D. Phillip Sponenberg (Wiley-Blackwell, 2009); also, the *First Edition* (Iowa State University Press, 1996); www.horse-genetics.com, by Glynis Scott, Ph.D.; and http://www.chboa.com by Champagne Horse Breeders & Owners Association. Both websites were visited October 16, 2009.

## The Eyes of the 7" Series Tennessee Walkers

Among the six families and four individuals, the eyes were painted in a variety of styles, but the basic eye color was always black. They are described here in chronological order.

On the Champagne Family, the greater eye area was sprayed black, and a white highlight was added to the black iris-pupil area. The highlight is a tiny dot representing reflection off the eye. It gives the eye an appearance of looking in a certain direction. There is a little black paint outside the eyeball itself, which is realistic for horses in general (but not for champagne colors, which have pink skin). Most horses have gray or dark skin, and it will show at the eyelids and slightly beyond, where the hair is thin.

On the Black Pinto stallion, the eyes are black with a white highlight, and a gloss coat was added, so the eyes are distinct from the black body color.

The Black (Pearled) Family has black eyes with a highlight. The eyes were not glossed, so their black color matches the adjacent body color.

The Black Pinto and Bay Pinto Foals both have a black eye with white hand-painted in the back corner (by brush). There is no highlight.

On the Buckskin Family, the eyes are sprayed black. They have a highlight, but no white corner and no gloss.

The Dapple Grey Stallion's eyes are black with a white highlight and a white back corner.

In the Bay Family, the foal's eyes are black with a white highlight, but his parents' eyes are black with a white back corner, but no highlight. I think the general eye area was sprayed black, then the eyes were painted white, and then black paint was sprayed over the white to make the black iris-pupil. The stallion and mare have a white back corner because it was deliberately left exposed. On models in general, there are two ways to get a white corner on a black eye: adding a little white to a black eye or adding black over a white eye and leaving some of the white showing!

In the Bay Roan Family, the first step was spraying the general eye area black. Then, on the foal's eyes, peach color was hand-brushed into the back corner. On the stallion and mare, peach or white was brushed into both the front and back corner of each eye. On the Bay Roan Family, the eyes do not have a highlight, but they were glossed.

In the Chestnut Roan Family, the general eye area was first sprayed black. All three family members have white added in the front and back corners of each eye, and a gloss coat over the top (but no highlight).

A lot of detailed painting work went into just the eyes on these models, and we're talking about horses under 7" tall!

**Extended Family**. After 2000, Hartland extended the 7" Tennessee Walker family by 22 models: six three-member families, two uncles, and two adopted foals! The largest addition, with 150 sets, was the Champagne Family. In second place, with 125 pieces, was the Black Pinto 7" Tennessee Walker Stallion. Remarkable for their graceful poses and delicate sculpture, the 7" series Tennessee Walker Family is the only horse family revived by Hartland 2000.

| Tennessee Walker Families: 7" Series – Production Models<br>Three-Piece Sets (Stallion, Mare, and Foal) | | | | | |
|---|---|---|---|---|---|
| **Color** | | **Item #** | **Run Type** | **Debut Date** | **Qty.** |
| Champagne (Pearled) | 3 pc Family | 684-01 | r | March 2001 | 150 |
| Black (Dappled & Pearled) | 3 pc Family | 684-02 | r | July 2001 | 50 |
| Buckskin | 3 pc Family | 684-03 | r | January 2002 | 50 |
| Bay (w/ Pearled Undercoat) | 3 pc Family | 684-04 | r | August 2003 | 25 |
| Bay Roan | 3 pc Family | 684-BR | r | May 2006 | 25 |
| Chestnut Roan | 3 pc Family | 684-CR | r | May 2006 | 25 |

**Family Table –** The production editions of 7" series Tennessee Walkers included six families, all regular runs released in 2001-2006. Three of the six families were in at least partly pearled colors. The largest edition, at 150 sets, was the Champagne Family.

| 7" Tennessee Walkers – Production Models<br>Sold Individually | | | | | |
|---|---|---|---|---|---|
| **Color** | **Individual** | **Item #** | **Run Type** | **Debut Date** | **Qty.** |
| Black Pinto | Stallion | 684S-01 | r | April 2001 | 125 |
| Black Pinto | Foal | 684F-03 | r | February 2002 | 50 |
| Bay Pinto | Foal | 684F-02 | r | February 2002 | 50 |
| Dapple Grey | Stallion | 684S-03 | r | September 2002 | 50 |

**Stallion & Foal Table –** Among 7" Tennessee Walkers, three pintos and a dapple grey were sold individually. The two stallions and two foals were regular runs released in 2001-2002. None had pearled color. Among the four, the largest edition, at 125, was the Black Pinto 7" Stallion.

# Chapter 5
# Arabians (and Part-Arabs)
## 11" Series

*Neighs rang out from Arabian stallions standing under the shade trees.* This sturdy stallion with proud demeanor is the largest size of Hartland Arabian. A member of the 11" series, he is 9.75" tall and 11.5" long. He's a good representation of an Arabian crossed with a heavier breed, without loss of quality.

**Arabians in 12 colors.** Details follow for the 12 production colors.

**A**
Buckskin Grey 11" Arabian, 901-06 – This example has beige pasterns; others have white pasterns (low socks).

**B**
Dapple Grey (Pearled) 11" Arabian, "Steadfast Hart," 2003 Club Model, 901-11.

**C**
Silver-Gray (Pearled) 11" Arabian, 901-02. Here, he photographed as very pale gray.

**D**
This photo of another Silver-Gray (Pearled) 11" Arabian, 901-02, captures the silver-gray coat at its most reflective.

**E**
The Black (Pearled) 11" Arabian with White Socks, 901-03, *at right* looks evenly black while the one *at left* appears to have a slightly lighter body.

**F**
This photo of a Black (Pearled) 11" Arabian with White Socks, 901-03, captures his pearled sheen.

## A

- Buckskin Grey 11" Arabian, 901-06; October 2001, 50 made; VG: $27... EX: $36... NM: $45.

This model's body is a pale, yellow-beige with a light gray undertone. It has black points with two exceptions: the lower part of the tail is pale beige and white, and the pasterns are either beige or white. The white pasterns ("socklets") are the unpainted white plastic. On an example with beige pasterns (illustrated), the beige color probably sprayed a little farther down than intended. The hooves are dark gray. The model is neither dappled, nor overtly pearled, but its body color can appear slightly luminous.

The light tail tip in the presence of black points confirms that it's a grey. The dark shading on the sides of the face go more with dun or grulla than buckskin, but there's no dorsal stripe (unless, if possible, it had already greyed away). The company called the color: rose grey. Collector Marla Phillips calls it "light rose grey," but I prefer "buckskin grey" because is more specific.

This model was an edition of 50 in October 2001, for $36. It was sold out by January 2003. *Photo A, courtesy of Marla Phillips.*

## B

- Dapple Grey (Pearled) 11" Arabian, "Steadfast Hart," 901-11; January 2003, 72 made; VG: $27... EX: $36... NM: $45.

"Steadfast Hart" was the club model in the $50 (ppd.) membership for 2003, which also included a lapel pin with his picture on it. The total number produced was 72. The company called him: pearl dapple grey.

"Steadfast Hart" has a metallic, silver-pearled body with black shadings on the body and dapples in the dark shadings. The mane, tail, muzzle, knees, hocks, and fetlocks are black, but the color inside his ears looks dark gray. None of the black color on the model is pearled. He has a right hind stocking and peach hoof below it; the other three feet are black. (In comparison, the Silver-Grey Arab – photos C & D – does not have dapples.)

## C & D

- Silver-Gray (Pearled) 11" Arabian, 901-02; June 2001, 50 made; VG: $27... EX: $36... NM: $45.

This model has a silver-gray, pearled body with pale shadings, black points, and white stockings in front. The black areas on his legs were not masked. (The edges are not crisp; they blend.) The areas not pearled are his peach front hooves, and the black regions: his mane, tail, rear lower legs, knees, and hocks. He does not have any dapples. The company called him: silver grey.

An edition of 50, the Silver-Gray Arabian debuted at Jamboree, June 22-23, 2001, as a surprise special run. He moved to the company website, and was sold out by mid-July 2001. His price was $36. *Photo C, courtesy of Eleanor Harvey.*

## E & F

- Black (Pearled) w/ White Socks 11" Arabian, 901-03; June 2001, 100 made; VG: $27... EX: $36... NM: $45.

The Black Arabian was another model that debuted at Jamboree in June 2001 as a surprise special run. He began as an unlimited edition for $36. The price changed to $32 in January 2003. (From November 15-December 20, 2002, the price was $27.) The edition size was capped at 100 by April 2003. There were 19 left in August 2004, and seven remaining in March 2007. They were all sold by February 2009. The company called his color: black pearl.

This Arabian's black body is pearled, and his four white socks and peach hooves are pearled. At least, that's how mine looks. His mane, tail, and muzzle, which are black, were not pearled though. He does not have any white on his face.

Two of the three Black Arabians illustrated are deeply and evenly black, but one has darker points than its body. Its owner, Marla Phillips, said the body has a distinct blue tint. Mine (not shown) is evenly black. *Photo E, courtesy of Marla Phillips. Photo F, courtesy of Eleanor Harvey.*

## G, H, & J

• Bay with Silver Dapples 11" Arabian, "Silver Sultan," 901-01; May 2001, 305 made; VG: $27... EX: $36... NM: $45.

"Silver Sultan," was the 2001 show special model, and the first 11" Arabian made since the 1990s. He was available at model horse shows that joined the company's Hartland Horses Show Producers Club for 2001. The shows were required to follow various rules, including charging no more than $45 for "Silver Sultan." There were usually some models left after the shows, and if contacted, show holders were happy to sell them by mail (for $45 plus shipping).

After 2001, "Silver Sultan" was sold through the company website and mailings. In August 2004, 20 were still available; in January 2007, six were left. "Silver Sultan" ranged in color from light to dark. A total of 305 were produced. Read more on him later in this chapter. *Photo H, courtesy of Eleanor Harvey.*

## I

• Flaxen, Liver Chestnut 11" Arabian, 901-09; April 2002, 50 made; VG: $30... EX: $40... NM: $50.

This model has a dark brown body and pale, golden beige (flaxen) mane and tail; high stockings (unpainted white plastic) on the hind legs; peach hooves below the stockings; and black front hooves, muzzle, and ear insides. There is also darker shading around the eyes and on the side of the face between eye and nostril. The company called him: liver chestnut.

The Flaxen, Liver Chestnut 11" Arabian was a run of 50 in April 2002. His original price was $36, reduced to $32 in January 2003. (The price was briefly $27 from November 15-December 20, 2002.) In August 2004, 14 were left. All were sold by May 2006.

**G**
This is a lighter version of "Silver Sultan," the 11" Arabian in Bay with Silver Dapples, the 2001 Show Special Model, 901-01.

**H**
This is a darker version of "Silver Sultan," the 11" Arabian in Bay with Silver Dapples, 901-01.

**I**
Flaxen, Liver Chestnut 11" Arabian, 901-09.

**J**
There really is a difference between lighter and darker versions of "Silver Sultan."

## K

- Copper Chestnut 11" Arabian, 901-04; July 2001, 50 made; VG: $30... EX: $40... NM: $50.

The Copper Chestnut Arabian has a burnt orange body with dark brown (near-black) mane, tail, and muzzle. His dorsal area, knees, and hocks are slightly darker than the body. His hind hooves are dark gray while the front hooves are peach. His two front socks are glossy white (unpainted plastic), and his tail tip is white. A hand-painted, white star peaks out from under his forelock. The body color glows, so might have a pearled undercoat. The company called him: copper chestnut.

This Arabian went up for sale in July 2001 for $36. All 50 of him were sold out by January 2002.

## L

- Dark Blue w/ White Mane & Tail, 11" Arabian, "Hart of Blues," 901-HB; August 2005, 30 made; VG: $27... EX: $36... NM: $45.

This Arabian has a dark, gray-blue body with black shadings and black lower legs and hooves. The white mane and tail were painted by hand.

The color was, no doubt, inspired by Hartland's 1960s, #876 blue, 9" grazing Arabian mares.

"Hart of Blues" was a run of 30 sold by a mail-in drawing through the company newsletter. Readers had about a two-week window in which to return an entry form. The 30 winners were notified in late July 2005, and paid $40 plus $6.95 shipping. The model arrived in August 2005.

(The company's catalog number, printed in its newsletter, was: 901VB, without a hyphen. Most likely, the "V" was a keyboarding error, and the model's number was meant to be 901-HB, for "Hart of Blues.")

**K**
Copper Chestnut 11" Arabian, 901-04.

**L**
Dark Blue with White Mane and Tail, 11" Arabian, "Hart of Blues," 901-HB.

## M & N

- Chestnut Pinto 11" Arabian, "Copper King," 901-07; Early 2002, 76 made: VG: $30... EX: $40... NM: $50.

"Copper King" was the show special model for the first half of 2002. His total run size was 76. He debuted at the Jamboree Benefit Show, February 9-10, 2002, in Pomona, California. The model in the photos came from Robyn Porter and Melissa Clegg's Northwest Congress in Washington state, May 18-19, 2002. They sold him for $45 plus postage. "Copper King" was also sold at shows held by Debra Kerr in Florida and Betsy Andrews in Maryland.

"Copper King" is a tobiano pinto with an orangey, light brown body color (similar to a light red dun). He has a brownish-orange mane and tail, black or dark brown shading on his muzzle, and dark brown color inside his ears. His four white stockings are glossy (unpainted). His tan areas look pearled, but his peach hooves are not.

## O & P

- Bay Pinto 11" Arabian, "Bronze Ruler," 901-08; Mid-2002, 46 made; VG: $27... EX: $36... NM: $45.

"Bronze Ruler" was the show special model for the second half of 2002, and was sold at Jamboree. There were fewer shows that year, and in December 2002, he was added to the company website (for $33.75 plus shipping). Then, he was $38 on mail order forms that arrived in late Jan. 2003. In April 2003, seven were left; they were gone by May 2004. The total run size was 46.

"Bronze Ruler" has a caramel-colored body (similar to light red dun), with black points except for four high, white stockings and white in the mane. His body color is somewhat reflective, but not obviously pearled. The white areas are glossy, unpainted plastic, and have crisp edges. (They were masked.) The peach hooves are not pearled.

**M**
The Chestnut Pinto 11" Arabian, "Copper King," 901-07, was the Show Special for the 1st half of 2002.

**N**
Chestnut Pinto 11" Arabian, "Copper King" – *left side.*

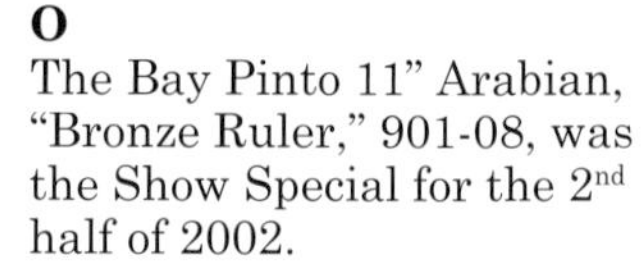

**O**
The Bay Pinto 11" Arabian, "Bronze Ruler," 901-08, was the Show Special for the 2nd half of 2002.

"Bronze Ruler" has the same pinto pattern as the chestnut pinto, "Copper King," but is a little darker shade of golden brown on the examples in the pictures. Still, he may have been created by adding black points to "Copper King."

## Q, R, & V

- Black Pinto w/ Star Face, 11" Arabian, 901-10W; December 2002, 15 with star; VG: $30… EX: $40… NM: $50.

A total of 50 Black Pinto Arabians were made, and 35 were given as gifts to volunteers at the 2002 Jamboree (August 16-18, 2002). They had a black forehead. Then, a (white) star was painted on the remaining 15, and they were sold through a drawing in the Hartland company newsletter.

Subscribers had about five weeks to enter by mailing their contact information in time for the November 15, 2002, deadline. The 15 winners paid $36 (plus $6.95 shipping). The model was shipped in December 2002. His black areas are normal black, not pearled, and his white areas are glossy white (unpainted). His tobiano pinto pattern matches the chestnut and bay pintos from earlier in 2002. He has peach hooves and four white stockings. *Photo V (face picture): model, courtesy of Dee Gwilt; photo, courtesy of Robyn Porter.*

**P**
Bay Pinto 11" Arabian, "Bronze Ruler" – *left side.*

**Q**
The Black Pinto with Star Face 11" Arabian, 901-10W, was sold; a version with no star was a gift run.

**R**
Black Pinto with Star Face 11" Arabian – *left side.*

## S & T

- Palomino (Dappled) 11" Arabian, 901-05; October 2001, 56 made; VG: $36… EX: $48… NM: $60.

This model is a medium palomino with an elaborate overlay of pale, rosy brown shading embedded with lighter dapples. There's light brown shading on the knees and hocks, and dark brown on the muzzle and inside the ears. The hind legs have high white stockings (unpainted plastic) and peach hooves while the front hooves are dark gray below half-pasterns (very low socks). White paint was added to the mane and tail. The company called the color: dapple palomino.

The palomino debuted in October 2001, for $36, and sold out by mid-January 2002. Meant to be a run of 50, "overstock" raised the total to 56.

## MORE ON "SILVER SULTAN," THE BAY ARABIAN WITH SILVER DAPPLES

"Silver Sultan," the 11" series Arabian in dappled bay, varied in color, even at his first show, which was Elaine Boardway's Salt City Sizzler in Syracuse, New York, on May 19, 2001. After the show, Elaine wrote that she kept "a definite bay" for herself. She offered me a "smoky looking guy," and others. I bought the light, warm bay in photos G & J.

**Lighter Version.** My Lighter Version of "Silver Sultan" (photos G & J) has a warm brown body with lots of silver-brown dapples on the head, neck, body, and upper legs. The body glows copper-colored. He has a black mane, tail, knees, hocks, muzzle, and groin, and a wide swath of black in the dorsal area.

**S**
The Palomino (Dappled) 11" Arabian, 901-05 is exactly this shade of yellow, with rosy brown shadings on the knees and hocks.

**T**
Palomino (Dappled) 11" Arabian, 901-05. The dapples show up better in this photo.

**U**
From any angle, the Black Pinto and Bay Pinto 11" Arabians have the same marking pattern; so does the Chestnut Pinto *(not shown)*.

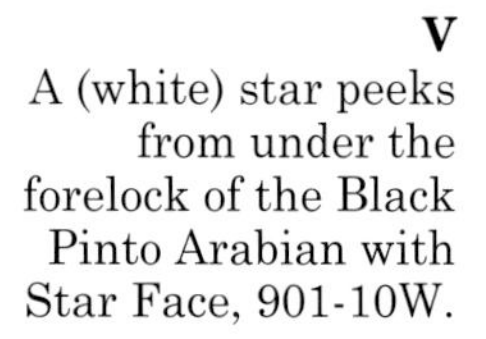

**V**
A (white) star peeks from under the forelock of the Black Pinto Arabian with Star Face, 901-10W.

**W**
Three of the 11" Arabians have a (white) star: the Copper Chestnut (901-04), Black Pinto with Star Face (901-10W), and Liver Chestnut (901-09).

The black areas are not pearled. He has four pearled white stockings, and peach hooves that are not pearled.

**At a Show Later in the Year.** Lauri Barnwell reported three variations of "Silver Sultan" at her show, Victorian Village Live, in Ferndale (northern), California, November 3, 2001. She described them as:

1. "A medium gold/silver with bright dapples and...satin shine,"
2. "Light, very bright gold/silver with tons of dapples" (which she said was less common),
3. "Very dark, almost a [dark] buckskin/bay with faint dapples and very matte."

My remark that the body on my Lighter Version (photo G, from New York), "glows copper-colored" is along the lines of Lauri's description, but where she wrote, "gold," substitute "copper." Brown paint over or beside silver paint yields the look of a warm, metallic color, but more like copper than gold, I think.

**Darker Version.** From Lauri, I bought a dark model (type 3). My dark version (in photo J) has the same color features as my light version (in G & J), with three exceptions:

1. His body is a shade darker (red-brown, instead of burnt orange),
2. His face has black shading,
3. More of his silver dapples have brown paint over the top of them.

Also see photo H, Eleanor Harvey's dark version.

**Labeling "Silver Sultan."** "Silver Sultan" is a hard-to-peg fantasy color. The company called him, "dapple rose grey," but that was confusing because the two other models it called "rose grey" look nothing like him. (The October 2001 Arabian the company called "rose grey" has a pale, beige-based body, and the February 2002 Five-Gaiter it called "dapple rose grey" has a white-based body with bay shadings.)

In my publication, *Hartland Model Equestrian News*, I labeled "Silver Sultan," as "metallic dapple dark rose grey" (September-October 2004), and then, "dappled dark bay—pearled" (November-December 2004). However, many examples are more like light or medium bay than dark bay. I now think he's best covered by "Bay with Silver Dapples."

**Ten Shows.** "Silver Sultan" was sold at 10 shows, all in 2001. They were:

1. Elaine Boardway's show in New York state, May 19
2. Jamboree in Pomona, California, June 22-23
3. Southern California's RazzMatazz Model Horse Show, August 25-26, Lake San Marcos, California
4. Cassie Hayes' Indian Nations Live show in Dewey, Oklahoma, Sept. 8
5. The Jamboree Benefit show Sept. 22 in Yucaipa, California
6. Pam Stamps' TMHCA's 2001 Octoberfest Live at The Saddle Rack in Elkmont, Alabama, Oct. 13
7. Karen Meekma's show (Dodge County Horseman's Assoc. / Crystal Creek Riders) Oct. 20 in Randolph, Wisconsin
8. Shay Canfield's Break A Leg Live Show, Oct. 20-21 at the National Guard Armory in Warrenton, Virginia
9. LeeAnne Masters' Horse Styles Fall Model Horse Show, November 3, Valley Center, California
10. Lauri Barnwell's show, at the Humboldt County Fairgrounds, Ferndale, California, November 3

Sources for the show information included: *The Hobby Horse News Special* (BreyerFest) *Edition*, July 2001; *Hartland Courier*, Summer 2001; Haynet message #80006 at www.yahoo.com, from Hartland, August 29, 2001; a Horse-Power Graphics flyer that arrived 9-17-01; private communication with Hartland (an email on September 21, 2001); and private communications from several collectors.

| Arabians (and Part-Arabs): 11" Series - Production Models | | | | |
|---|---|---|---|---|
| **Color & Name** | **Item #** | **Run Type** | **Debut Date** | **Qty.** |
| Bay with Silver Dapples, "Silver Sultan" | 901-01 | Show special, 2001 | May 2001 | 305 |
| Silver-Gray (Pearled) | 901-02 | r | June 2001 | 50 |
| Black (Pearled) with White Socks | 901-03 | r | June 2001 | 100 |
| Copper Chestnut | 901-04 | r | July 2001 | 50 |
| Palomino (Dappled) | 901-05 | r | Oct. 2001 | 56 |
| Buckskin Grey | 901-06 | r | Oct. 2001 | 50 |
| Chestnut Pinto, "Copper King" | 901-07 | Show special, 1st half of 2002 | Feb. 2002 | 76 |
| Flaxen, Liver Chestnut | 901-09 | r | April 2002 | 50 |
| Bay Pinto, "Bronze Ruler" | 901-08 | Show special, 2nd half of 2002 | Mid-2002 | 46 |
| Black Pinto with Star Face | 901-10W | Sold to winners of mail-in drawing | Dec. 2002 | 15 |
| Dapple Grey (Pearled), "Steadfast Hart" | 901-11 | Club model, 2003 | Jan. 2003 | 72 |
| Dark Blue with White Mane and Tail, "Hart of Blues" | 901-HB | Sold to winners of mail-in drawing | Aug. 2005 | 30 |

**Arabians** – Of the 12 production colors of 11" series Arabians, six were regular runs, and six were special runs. "Silver Sultan," the 2001 show special model, was not only the most numerous Arabian, but also the best selling horse from Hartland 2000 (not counting the two horse-and-rider sets). The comparison may not be fair since most models were limited editions. Who knows how many of each could have sold if given the chance?

**Majestic.** The 11" series Arabian is a signature sculpture from the Hartland line. "Majestic" describes this stallion keeping watch over his own.

## Chapter 6

# Quarter Horses and More

## 11" Series

*Quarter Horses, Appaloosas, and Paints snorted and pranced in the paddock.* With head tucked and neck arched down, the 11" series Quarter Horse is 11.25" long, and 7.75" tall at the highest part of the neck. His prancing pose comes closest to trotting, but interpretations of backing or cantering are reasonable.

**A**
Buckskin (Dappled) 11" Quarter Horse, "Viceroy," Noble Horse Series, 902-11.

**B**
The Dapple Buckskin 11" Quarter Horse has beige guard hairs at the top of the tail and stripes on his hooves.

**C**
Palomino (Dappled) 11" Quarter Horse, 902-04. The dapples are most apparent on the neck and shoulders.

**D**
Dapple Palomino Quarter Horse, 902-04 – *left side.* This photo exactly matches his light palomino color.

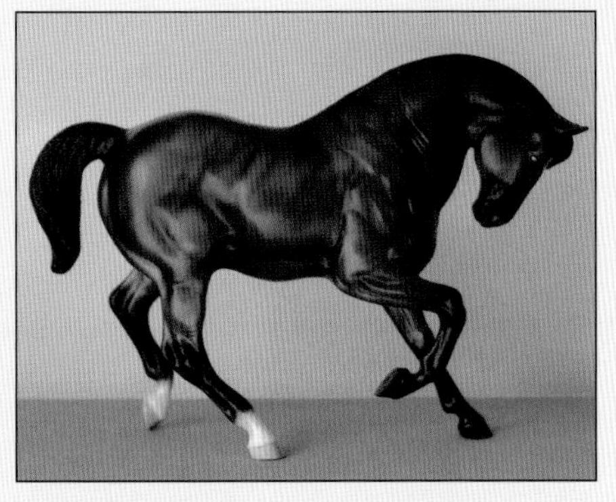

**E**
Chocolate Bay 11" Quarter Horse, 902-02.

**F**
Chocolate Bay 11" Quarter Horse – *left side.* This mold has rippling shoulder muscles.

**Nine colors.** The 11" series Quarter Horse mold can represent any stock breed or stock-draft cross. Dramatically muscled, this model could be a gelding (or unusually sturdy mare). Another distinctive feature is his roached (crew cut) mane. Nine production colors were painted on the 11" Quarter Horse mold: four solid (non-spotted) colors, two pintos, and three appaloosas.

## A & B

- Buckskin (Dappled) 11" Quarter Horse, "Viceroy," 902-11; June 2006, 20 made; VG: $60... EX: $80... NM: $100.

"Viceroy" was the first model (of two) in the Noble Horse Series. Each of the 20 "Viceroy" models was painted by Hartland's paint-scheme designer, Sheryl Leisure, who autographed each model's underside.

"Viceroy" has a golden, yellowish-tan body with subtle, large dapples; black muzzle; and black points except for tan guard hairs at the top of the tail and along the edge of his (roached) mane. His hooves match his body except for faint, vertical striping. He has a dark dorsal stripe, and a masked, large star (or short blaze) in white paint. The company called him "dapple buckskin," and said that he was "hand dappled."

"Viceroy" was sold by a mail-in drawing through the company newsletter that arrived in early June 2006. Subscribers mailed their contact information on a 3 x 5" card or postcard. The 20 winners paid $75 plus $6.95 shipping.

"Viceroy" came with a certificate of authenticity with his picture on it, and his picture also appeared on a full-color card stapled to the company newsletter issue that advertised him. (It was the Winter 2006 issue.) "Viceroy" has an edition number written on his underside; for example: 19/20.

## C, D, & G

- Palomino (Dappled) 11" Quarter Horse, 902-04; February 2002, 50 made; VG: $36... EX: $48... NM: $60.

This palomino has a pale-yellow body and dapples framed by pale, reddish-gold shadings; white mane and tail, four white socks, four peach hooves, and a dark muzzle. The boundaries of the white socks were freehand-airbrushed, rather than masked. The white socks are the bare plastic.

This model was released in February 2002 for $38. By January 2003, all 50 had been sold.

## E & F

- Chocolate Bay 11" Quarter Horse, 902-02; October 2001, 50 made; VG: $33... EX: $44... NM: $55.

This model is dark brown with black points, except for hind socks. The socks look pearled; the rest of the model is semi-glossy, but not pearled. The hind hooves are peach while the front hooves are black. This bay's face has a large, white star and stripe and a large pink snip.

This model appeared at the website in September 2001, and was shipped in October 2001. The 50 pieces were sold out by mid-January 2002. I believe the price was $36. The company called it: chocolate bay.

## H

- Bright Bay 11" Quarter Horse, 902-09; July 2002, 50 made; VG: $36... EX: $48... NM: $60.

This model has a matte, burnt orange body color; black muzzle, mane, tail, knees, and hocks; four white stockings, and peach hooves, including the bottoms of the two raised hooves. The stockings were masked and are the unpainted, white plastic. A masked blaze, which is painted on, ends in peach at the nostrils.

The company called this model "shaded red bay," but it does not have shadings, and some Quarter Horses from previous Hartland companies had truly red bodies, so "bright bay" is a better term.

The Bright Bay went up for sale in late June 2002, and arrived in July. He was $36. All 50 pieces were sold out by November 2002.

## I & J

- Black, Tobiano Pinto / Paint, 11" QH mold, "Tom-Tom," 902-01; June 2001, 250 made. VG: $33... EX: $44... NM: $55.

**G**
The Palomino Quarter Horse, 902-04 *(left)*, is a paler yellow than the Palomino Arabian, 901-05. Both are dappled.

**H**
Bright Bay 11" Quarter Horse, 902-09.

**I**
Black, Tobiano Pinto / Paint, 11" Quarter Horse mold, "Tom-Tom," 902-01.

**J**
"Tom-Tom," the Black, Tobiano Pinto / Paint has high white stockings (hip waders) and is nearly 50% white.

**K**
Slate-Gray Overo Pinto / Paint (11" Quarter Horse mold), 902-08.

**L**
Slate-Gray Overo Pinto / Paint – *left side.*

The black-and-white Paint, "Tom-Tom" was the ticketed, Hartland special run for Jamboree 2001 (June 22-23, 2001). Collectors pre-ordered him, paying $50 for a ticket that had to be presented when picking up the model at the event. Extras were made: The total run was 250. In 2002, "Tom-Tom" was still available from the Jamboree host, Horse-Power Graphics, Inc., by mail, for $60 plus $6 shipping.

"Tom-Tom" is a black-and-white tobiano pinto (about 50% white), with four white stockings and peach hooves. His white areas are glossy, unpainted white except for his blaze, which is painted on in white; it turns peach-colored between the nostrils. The model's finish was not pearled. *Photos I & J, courtesy of Eleanor Harvey.*

## K & L

- Slate-Gray Overo Pinto / Paint, 11" QH mold, 902-08; June 2002, 50 made; VG: $33... EX: $44... NM: $55.

This horse has a dark- to medium-gray body with masked white areas, forming an overo pinto pattern with a white face, a white tail tip, and two hind stockings, with peach feet below them. The white areas are glossy, unpainted white plastic, except for the bald (white) face marking, which was painted on and turns to peach at the muzzle.

The last day of May, 2002, the company sent a postcard announcing that this "slate grey overo pinto," a 50-piece edition, would be available June 6 (2002) for $38. All were sold by September 2002.

## M & N

- Black Appaloosa with Dusty Hips, 11" QH mold, 902-10; January 2003, 50 made; VG: $36... EX: $48... NM: $60.

At first glance, this is a black horse with four white stockings and peach hooves, but it has a light dusting over the hips: a lattice of white that frames many small black spots and a few large ones. The white socks have crisp demarcations. A masked blaze ends in peach color near the nostrils. The company called it a "black blanket appaloosa" or "a black lacy blanket appaloosa."

This model debuted in January 2003; its price was $34.50. In April 2003, 12 pieces were left out of 50. He appeared on the August 2003 catalog sheet, and was sold out by May 2004.

## O & P

- Blue Roan Appaloosa, 11" QH mold, 902-07; December 2001, 100 made; VG: $36... EX: $48... NM: $60.

This Appaloosa has a light gray body with black head and points except for hind socks. The socks are unpainted, white plastic. It has a white hip blanket with black spots. Each hind leg has two ermine spots on the coronet band, and brown stripes on the otherwise peach-colored hooves. The front hooves are black. The model's face has a masked, narrow white blaze painted on.

This model came out in December 2001 as an edition of 100 for $38. (During November 15-December 20, 2002, it was $28.50.) In January 2003, when the price was raised to $34, only a few were left. They were all sold by the end of January 2003.

## Q & R

- Buckskin Appaloosa, 11" QH mold, 902-03; eBay (online) special run, September 2001, 100 made; VG: $39... EX: $52... NM: $65.

The Buckskin Appaloosa was an eBay Store Special Edition, available from Hartland only through the Internet auction giant that began doing business in 1995. The Appaloosa went up for sale on August 28, 2001, and 30 out of 100 were sold by September 3. The price was $45 plus $6.95 shipping.

This model has a very pale tan body color, a snowflake effect on his hips (like fine white dapples), and dark brown body spots inside and outside the white on the hips. There are dark brown leg bars (horizontal stripes on the upper legs). He has a black muzzle and black points (mane, tail, and lower legs) except for buckskin guard hairs on his tail and a right hind sock. The sock is unpainted white plastic.

His right hind hoof is peach except for two brown, vertical stripes. That leg also has a dark-brown ermine spot on the coronet band, above the hoof stripes. The other three hooves are black. His face has a large (white) star or short blaze painted on.

**M**
Black Appaloosa with Dusty Hips (11" QH mold), 902-10.

**N**
The Black Appaloosa with Dusty Hips has a lattice of white on its hips that frames many small black spots and a few large ones.

**O**
Blue Roan Appaloosa from the 11" Quarter Horse mold, 902-07.

**S**
The Chocolate Bay 11" Quarter Horse *(third from left)* has a white blaze and pink snip. Others with face markings are: the Black Pinto (902-01), Buckskin Appaloosa (902-03), and Blue Roan Appaloosa (902-07).

**P**
Blue Roan Appaloosa – *left side.* His main color is light gray; the white hip blanket has medium-sized black spots.

**T**
More 11" Quarter Horse mold models with face markings are: the Slate-Gray Overo Pinto (902-08), Bright Bay (902-09), Black Appaloosa (902-10), and Dappled Buckskin (902-11). Only the Palomino *(not shown)* has no white on its face.

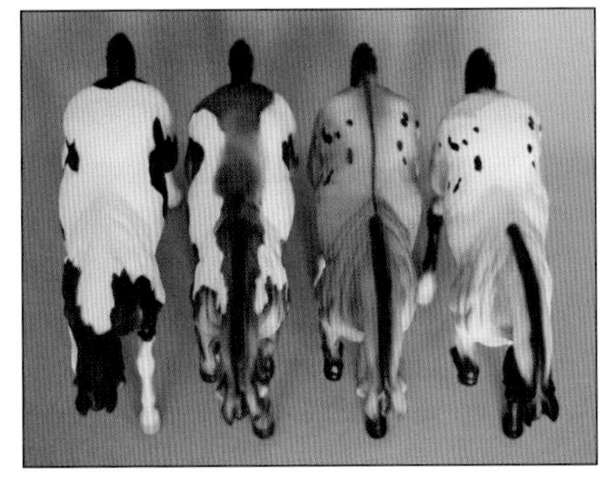

**U**
Spotted Horses, *from left* – The Black Pinto (902-01), Slate-Gray Overo Pinto (902-08), Buckskin Appaloosa (902-03), and Blue Roan Appaloosa (902-07). Both appaloosas have the same pattern of dark spots on the hips.

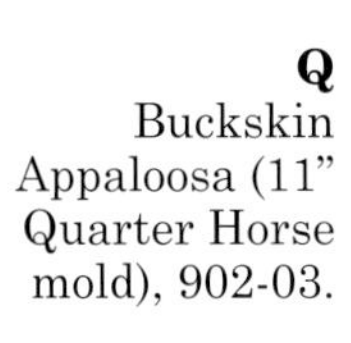

**Q**
Buckskin Appaloosa (11" Quarter Horse mold), 902-03.

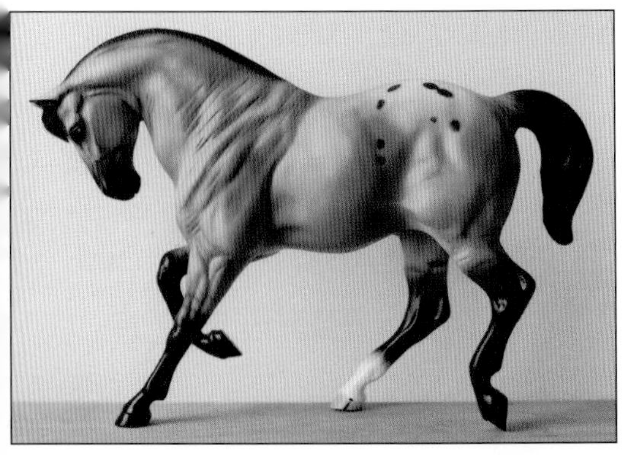

**R**
Buckskin Appaloosa, 902-03 – *left side.* He has brown leg bars, one striped hoof, and a white snowflake effect and dark brown spots over his hips.

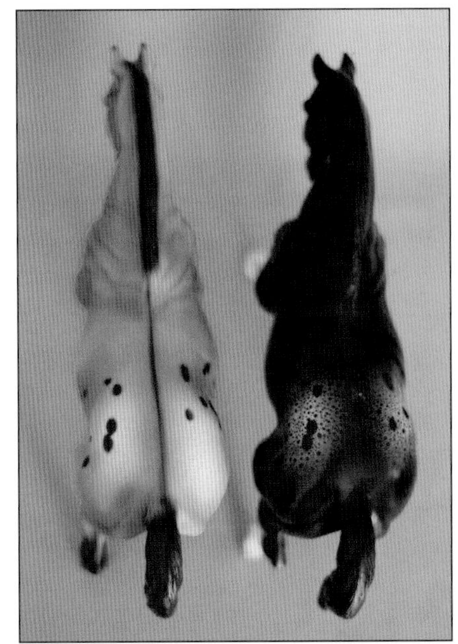

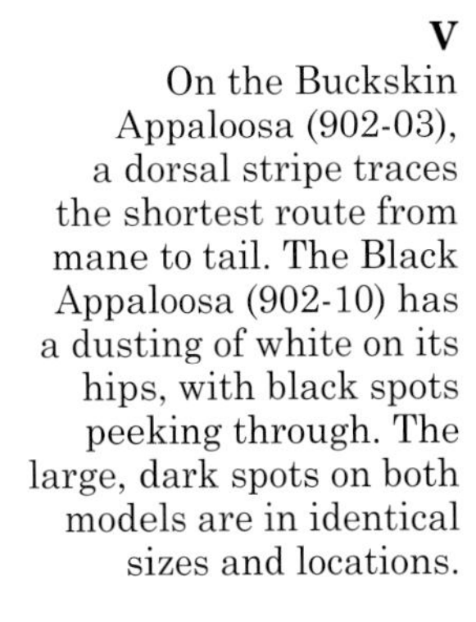

**V**
On the Buckskin Appaloosa (902-03), a dorsal stripe traces the shortest route from mane to tail. The Black Appaloosa (902-10) has a dusting of white on its hips, with black spots peeking through. The large, dark spots on both models are in identical sizes and locations.

| Prancing Quarter Horses, Appaloosas, and Paints: 11" Series Production Models | | | | |
|---|---|---|---|---|
| **Color & Name** | **Item #** | **Run Type** | **Debut Date** | **Qty.** |
| Black Tobiano Pinto / Paint, "Tom-Tom" | 902-01 | Ticketed Jamboree special, 2001 | June 2001 | 250 |
| Buckskin Appaloosa | 902-03 | eBay (online) special | Sept. 2001 | 100 |
| Chocolate Bay QH | 902-02 | r | Oct. 2001 | 50 |
| Blue Roan Appaloosa | 902-07 | r | Dec. 2001 | 100 |
| Palomino (Dappled) QH | 902-04 | r | Feb. 2002 | 50 |
| Slate-Gray Overo Pinto / Paint | 902-08 | r | June 2002 | 50 |
| Bright Bay QH | 902-09 | r | July 2002 | 50 |
| Black Appaloosa with Dusty Hips | 902-10 | r | Jan. 2003 | 50 |
| Buckskin (Dappled) QH, "Viceroy," 1st in "Noble Horse Series" | 902-11 | Sold to winners of mail-in drawing | June 2006 | 20 |

**Quarter Horses** –The head-down, prancing 11" Quarter Horse mold was painted in nine production colors, including three appaloosas, two pintos, and four solid colors. Six were regular runs, and three were special runs. Most often, they were editions of 50. The largest edition, at 250, was "Tom-Tom," the black pinto. He was also the third largest Hartland horse edition since 2000. (That doesn't count the two horse-and-rider sets.)

**Regal.** The 11" series Quarter Horse is one of the best action horse sculptures reproduced in plastic by any manufacturer since the beginning of plastic. It's truly regal.

# CHAPTER 7
# POLO PONIES
## 9" SERIES

*Polo Ponies, nimble and alert, dashed about at their riders' commands.* The Polo Pony is 8.25" tall, 9" long, and wears protective tail and leg wraps. He (or she) depicts a right-lead canter with a little bounce in it.

**Sixteen Colors.** Hartland 2000 released the Polo Ponies in a total of 16 production colors. Among the spotted horses, all five pintos have normal (non-pearled) finishes, but the two appaloosas are partly pearled. Hartland Polo Ponies have been made in only one size (scale), 9", so the size is not repeated in each model description.

**A**
Metallic Copper 9" Polo Pony, "Cinn," 2004 Club Model, 883-17.

**B**
Metallic Gold Bay 9" Polo Pony, "Heart of Gold," 2004 Holiday Horse, 883-16.

**C**
Flecked Grey 9" Polo Pony with Turquoise Wraps, 883-02. The color varied: This one has a near-white body.

**D**
Flecked Grey 9" Polo Pony with Turquoise Wraps, 883-02. This darker example has much shading.

**E**
Dapple Grey 9" Polo Pony with White Wraps, 883-07.

**F**
Cinnamon Dapple (Pearled) 9" Polo Pony, "Sugarplum," 2001 Holiday Horse, 883-06.

## A

- Metallic Copper Polo Pony, "Cinn," 883-17; 2004 Club Model, March 2004, 85 made; VG: $21... EX: $28... NM: $35.

This model's metallic copper body is very reflective. His white mane, tail, and wraps were painted creamy white (without being masked). The shiny black hooves were painted without a mask. There is some pale, darker shading on the body.

"Cinn" was one of two horses offered as the 2004 club membership model. For $50, you could get either "Cinn" or "Buckeye" (a semi-rearing horse from the rider series), a lapel pin picturing both horses, and the company newsletter. For $80, you received both models, but only one lapel pin. Some models remained after the membership year. On a November 2005 price list, each of the two club models, sold alone, was $55.

The company newsletter reported "Cinn" as a run of 50. Then, in February 2009, "Cinn" was for sale in the eBay "store" of Hartland distributor Horse-Power-Graphics, Inc., which said that 20 pieces were left (out of an unstated total). The price was $19.99. No lapel pin was included, and the models did not have the edition numbering. (Normally, the model's underside had a sequence number written on it; for example: #029.) In late December 2009, six of "Cinn" remained. They have since sold out.

In January 2012, Horse-Power Graphics' Sheryl leisure explained that 40 were sold as numbered, club membership models, and after that, the factory overrun of 45 more were sold unnumbered, for a total of 85 Metallic Copper Polo Ponies. As with each model, a few extras were ordered, but in this instance, there were dozens of extras!

## B

- Metallic Gold Bay Polo Pony, "Heart of Gold," 883-16; 2004 Holiday Horse, 85 made; VG: $21... EX: $28... NM: $35.

"Heart of Gold" has a metallic gold body; creamy white wraps that were brush-painted (and not masked); and black muzzle, hooves, and points except for the white bandages.

This model was the 2004 Holiday Horse, available from November 15, 2004 through January 31, 2005, but limited to 100 pieces. The price was $28 plus $6.95 shipping. It came with a sturdy, paper tag on a gold string, and was edition-numbered on the belly; for example: #40.

An edition size of 50 was reported in two company newsletters: Winter 2005, postmarked April 2005, and Spring/Summer/Fall 2006, sent in January 2007. In March 2007, the company website said four were left. Then in February 2009, "Heart of Gold" was for sale in Horse-Power Graphics' eBay store, which said that 20 were left out of 100 made. In December 2009, when the final two were for sale, the edition size was, again, reported as 100.

In January 2012, Horse-Power Graphics' Sheryl Leisure said that the sales records showed a total of 85 sold: about 65 during the 2004 holiday season, and about 20 more, factory overruns, sold later, unnumbered and without the hang tag.

In December 2009, with two pieces left, the eBay description read, "Hart of Gold." When the model was introduced in the Summer 2004 company newsletter (postmarked December 17, 2004), it was pictured with its tag. The tag may have been printed with an unintended spelling, "Heart," which was continued even years after the model was sold without the tag. Any way you spell it, though, Hartland models have "heart."

## C & D

- Flecked Grey Polo Pony w/ Turquoise Wraps, 883-02; June 2001, 50 made; VG: $18... EX: $24... NM: $30.

Two photos are shown because these flecked greys varied from nearly white-bodied (photo C) to dark gray (photo D), depending on how much dark shading was added. A Hartland employee, Denise, said that "most were a medium shade, but a few were very light, and a few were very dark." The run total was 50.

All are peppered with tiny, black specks simulating the dark hairs often found in the coat of grey horses. The knees and hocks are dark gray or black, and the hooves and muzzle are black. The black and gray areas are matte, but the turquoise wraps are glossy.

The Flecked Grey Polo Ponies with turquoise wraps debuted at Jamboree 2001 (June 22-23, 2001) as a surprise special run. The price was $32. After Jamboree, they moved to the company website, and were sold out by mid-July 2001. *Photo C (lighter version), courtesy of Eleanor Harvey.*

**Grey Horses and Grey Models.** The company called this model, 883-02, "flea-bit grey," but in real horses, that term usually denotes older greys with reddish brown hairs (specks) and, incidentally, no dapples.

In real-horse terminology, the darker version would be called, "iron grey." The lighter version would just be called "grey."

Calling it "flecked" gets across the idea that the model painter went to the effort of adding some colored hairs. Otherwise, (light) grey models usually are represented by a white body without flecks (but sometimes with shadings). A flecked body takes more effort to produce than a white body, and is not common in model horses.

## E

- Dapple Grey Polo Pony w/ White Wraps, 883-07; February 2002, 50 made; VG: $18... EX: $24... NM: $30.

This model has pale gray (near-white) dapples on a field of light gray. The mane, tail, knees, and hocks are black; the hooves are peach; and the white wraps appear to be unpainted (the bare plastic). This model's color is not pearled.

The original price was $34, lowered to $29 in January 2003. (From November 15-December 20, 2002, it was on sale for $25.50.) There was one left in mid-November 2004.

## F

- Cinnamon Dapple (Pearled) Polo Pony, "Sugarplum," 883-06; 2001 Holiday Horse, 150 made; VG: $24... EX: $32... NM: $40.

"Sugarplum" has pale copper shading over its body and dapples that are pearl (silvered) white. The mane and tail, and the legs from about the knees and hocks on down, are solid, pale copper color. The bandages on the legs and tail are pearl white. The hooves look like black over copper color. The muzzle, ear insides, and groin are black.

The company called the color "cinnamon dapple," "dapple cinnamon," or "cinnamon pearl dapple." It is a fantasy color. The darker areas on this model are similar to the body color on the "red roan" Hartland Five-Gaiter in styrene plastic, notably seen in a 1991 set from JCPenney. In *Hartland Horses and Dogs* (2000), I dubbed that model's color "cinnamon." For this Polo Pony, "cinnamon dapple" is a fair color description.

"Sugarplum," the 2001 Holiday Horse, became available in November 2001, and could only be ordered through January 31, 2002. His price of $36 included "Lump of Coal," a black Tinymite horse. "Sugarplum" was a run of 150.

**G**
Dun / Grulla 9" Polo Pony, 883-10.

## G

- Dun / Grulla Polo Pony, 883-10, December 2004, 39 made; VG: $15... EX: $20... NM: $25.

This horse falls in the dun / grulla color group. It is drab, flat brown with darker points and primitive marks: a dorsal stripe and horizontal, leg stripes. The body looks like it was light brown, but was mostly covered with darker, drab brown shadings over the top line, head, shoulders, and hips. The head is dark brown with some black on the forehead.

On this Polo Pony model, the browns have no reddish tone. The mane, tail, and hooves are black, but the pasterns are lighter (not solid black). The white bandages were painted on by hand without a mask. The dark brown, horizontal leg stripes zig zag somewhat. The company called the color: grullo.

This model became available in December 2004 for $19.99. It was called a "sample run" because it was a sample of the work by a different painting service. Four of the dun Polo Pony models were still left in March 2007.

**Head Color on Dun and Grulla Horses.** A Polo Pony released in 2004 was in the dun and grulla color group. On dun and grulla horses, at least part of the head – the bridge of the nose or the front half of the head, or maybe even most of the head – will be darker than the body color. If the darker color is black, that points to grulla; if it's some shade of brown, it could be a grulla or a dun, but more often, a dun. In contrast, the head of a buckskin horse will usually be about the same color as the body. That is what I understand from reading Sponenberg (*Equine Color Genetics*, 1996 and 2009).

It used to be that the "grulla" label required a grayish body, but under Sponenberg's categories, the body of a grulla can also be beige or light brown like some of the dun or buckskin colors.

On roan horses – bay roan, blue roan, etc. – usually the entire head will be darker than the body color. On dun and grulla horses, the dark color on the head is usually not that extensive.

**H**
Bay Roan 9" Polo Pony, 883-15.

# H

- Bay Roan Polo Pony, 883-15, December 2004, 44 made; VG: $18... EX: $24... NM: $30.

The Bay Roan Polo Pony has a body that is mostly off-white with some pale, rust-brown shadings. The head is dark brown, and the points are black except for the white bandages, which were painted freehand with a brush. The hooves, including the bottom of the raised hoof, are black. The model has a white star and stripe and white snip on its face.

This was another of the five "sample run" models that went up for sale in December 2004 for $19.99. Two Bay Roans were still left in January 2007.

The company newsletter stated that the quantity of this model and of the Bay Tobiano Pinto Polo Pony were both 44. That could well be true, but seems an odd coincidence.

**I**
Bay Tobiano Pinto 9" Polo Pony, 883-13.

# I & J

- Bay Tobiano Pinto Polo Pony, 883-13; December 2004, 44 made; VG: $18... EX: $24... NM: $30.

This tobiano pinto has a complex pattern of white markings, and was both spray-painted and brush-painted. It has a brown body with black shadings, and black on the mane, tail, and upper part of the muzzle. However, the mane is mostly white, and the muzzle is mostly pink.

The face has a joined star and stripe and a separate snip. The white markings on the face and body look spray-painted over the brown color, using masks. In some places, the edges of white areas look touched up with a brush.

The cranberry red wraps were brush-painted without a mask. The peach hooves were brush-painted, and are lighter than the hooves on most Hartland models painted since 2000. All hoof bottoms were entirely painted.

An alternate painting service did the work, so it was called a "sample run." Some models had small paint runs. The price was $19.99. This model came out in December 2004. The quantity made was 44. They were sold out by May 2006.

# K & L

- Liver-Chestnut Overo Pinto Polo Pony, 883-04; November 2001, 50 made; VG: $24... EX: $32... NM: $40.

This model is medium brown (liver chestnut) with a "flaxen" mane and tail that are actually the same color as the hooves: peachy-pink! The white, overo pinto markings are painted on, but the wraps and pasterns are unpainted white. The painted-on white blaze ends in peachy-pink between the nostrils.

This model debuted in November 2001 as an edition of 50 for $34 each. In mid-January 2002, eight were left. They were sold out by April 2002.

**J**
Bay Tobiano Pinto 9" Polo Pony – *left side.*

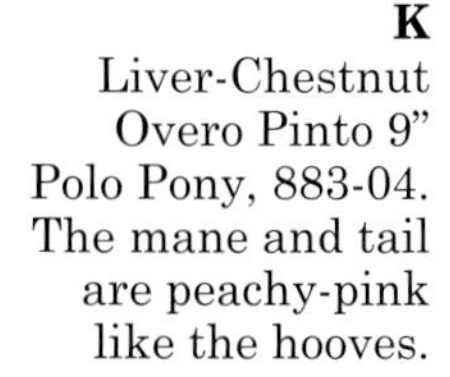

**K**
Liver-Chestnut Overo Pinto 9" Polo Pony, 883-04. The mane and tail are peachy-pink like the hooves.

**L**
Liver-Chestnut Overo Pinto 9" Polo Pony – *left side.*

## M & BB

- Pale Palomino Polo Pony w/ White Wraps, 883-05; September 2002, 35 made; VG: $24... EX: $32... NM: $40.

This palomino with white wraps has peach hooves, a black muzzle, black around the eye area, and a pale golden yellow body that is slightly darker where the airbrush was trained longer. The mane, tail, and stockings are the glossy, unpainted white plastic. Its white blaze was painted on, and turns pink between the nostrils.

In September 2002, the original price was $34, reduced to $29 in January 2003. (From November 15-December 20, 2002, the price was $25.50.) This model was advertised as an edition of 50. Twenty were left in April 2005, and in May of 2006, the company called it a sold-out run of 35.

**M**
Pale Palomino 9" Polo Pony, 883-05.

## N & BB

- Palomino (Dappled) Polo Pony w/ Green Wraps, "Volant," 883-19, Noble Horse Series; January 2007, 15 made; VG: $45... EX: $60... NM: $75.

This darker of the two palomino Polo Ponies, "Volant," has green wraps, and only 15 were made. A numbered edition (example: 09/15), it is a medium palomino with large dapples. There is only a subtle color difference between the dapples and the slightly darker, golden shading that frames them. The white color on the mane and tail looks painted on, as does the white face marking. The black shading around the muzzle and around the eyes indicates dark skin. Each nostril has a salmon pink "highlight" in it. The wraps are glossy, medium green.

"Volant" was the second of two horses in the Nobel Horse Series, which were models painted by Hartland designer Sheryl Leisure, rather than by general painting staff or outside painters. The model was announced in the company's final newsletter, which arrived in January 2007. Readers had about three weeks in which to submit their contact information on a 3" x 5" card or postcard. Fifteen names were randomly drawn. Winners paid $50 (plus $6.95 shipping).

**N**
Palomino (Dappled) 9" Polo Pony with Green Wraps, "Volant," Noble Horse Series, 883-19. This model was hand-dappled.

**O**
Buckskin Overo Pinto 9" Polo Pony, "Jammer Time," 2002 Jamboree Special, 883-08.

## O & P

- Buckskin Overo Pinto Polo Pony, "Jammer Time," 883-08; 2002 Jamboree Special; August 2002, 100 made; VG: $24... EX: $32... NM: $40.

Of the three overo pinto Polo Ponies, this one has the largest markings. The white areas, including the wraps, appear to be bare white plastic. Otherwise, the model is pale buckskin (beige) with black knees, hocks, pasterns, and hooves. He has an apron face, and peachy-pink muzzle.

"Jammer Time" was the ticketed, Hartland special run at Jamboree in August 2002, and was limited to 100 pieces. Those who bought a ticket in advance paid $36, and had to pick up the model at the show. For $41, anyone in attendance could buy one. After the event, Horse-Power Graphics, Inc., sold him by mail for $52 ppd. In June 2003, only four were left.

## Q & R

- Black Overo Pinto Polo Pony, "Black Tie," 883-09; 2003 Show Special Model; 78 made; VG: $24... EX: $32... NM: $40.

"Black Tie" is black with an apron face, peach muzzle, and overo pinto markings. The somewhat blurry edge on the body markings resembles the "mapping" effect on real pintos. The white wraps are glossy (unpainted), and their edges were masked. The white, apron face and pinto markings were painted on.

"Black Tie" was the 2003 show special model. Elaine Boardway sold him at her Salt City Sizzler model horse show in Baldwinsville, New York on June 14, 2003. After the show, she sold him at BreyerFest for $35, and by mail for $40 ppd. He was also sold at Jamboree, October 25-26, 2003, for $38. A total of 78 of "Black Tie" were made.

**P**
Buckskin Overo Pinto 9" Polo Pony, "Jammer Time" – *left side.*

**Q**
Black Overo Pinto 9" Polo Pony, "Black Tie," 2003 Show Special Model, 883-09.

**R**
Black Overo Pinto 9" Polo Pony, "Black Tie" – *left side.*

## S & T

- Black Appaloosa (Pearled) Polo Pony w/ Pale Gold Wraps, 883-03; July 2001, 75 made; VG: $27... EX: $36... NM: $45.

This appaloosa is black with a white hip blanket and black spots on the blanket. The white blanket is pearled, and so are the black areas on the body itself, but the knees, hocks, pasterns, and hooves are not pearled. On the black hip spots, some "silvering" from the pearled white blanket shows through. The wraps are the same pale gold / dark cream / ecru color as on the Copper Bay Appaloosa Polo Pony.

This model, which the company called, "black appaloosa pearl," debuted in July 2001 as an unlimited edition for $34. By December 2001, the edition was limited to 100. In January 2003, the price became $29. (From November 15-December 20, 2002, it was briefly $25.50.) Then, the edition size was capped at 75 by April 2003. One piece was left in November 2005.

**S**
Black Appaloosa 9" Polo Pony with Pale Gold Wraps, 883-03.

**T**
Black Appaloosa Polo Pony with Pale Gold Wraps – *left side.*

**U**
Copper-Bay Appaloosa Polo Pony with Pale Gold Wraps, "Chart the Course," 2001 Club Model, 883-01.

**V**
Copper-Bay Appaloosa Polo Pony w/ Pale Gold Wraps, "Chart the Course" – *left side.*

**W**
Red-Chestnut Overo Pinto 9" Polo Pony, 883-18. The pinto markings were a little different on each one.

**X**
Red-Chestnut Overo Pinto 9" Polo Pony – *left side.*

## U & V

- Copper-Bay Appaloosa Polo Pony w/ Pale Gold Wraps, "Chart the Course," 883-01; 2001 Club Model; 270 made; VG: $27… EX: $36… NM: $45.

"Chart the Course" was the club model in the $50 (ppd.) membership for 2001, which also included a lapel pin with his picture on it. The first ones were shipped in February 2001. A total of 270 were made. They are edition-numbered on the underside; for example: 101.

"Chart the Course" is a warm (copper brown) bay with a pearled white hip blanket with dark brown spots. In addition, there are pale, copper-brown dapples in the brown areas adjacent to the blanket (within an inch or more of it).

The blanket spots and the black areas on the model (mane, tail, lower legs, muzzle, and hooves) are not pearled. The copper brown body has a matte finish, but is faintly metallic, so probably has a pearled layer under it. The wraps, which are also faintly metallic, are glossy cream or pale gold; collector Pamela Pramuka calls them "ecru." The company called this model a "bay appaloosa," or just referred to him by name.

"Chart the Course" was the first model horse released by Hartland Collectibles, L.L.C. In *Hartland Model Equestrian News*, March-April 2001, I wrote that its color "is both attractive and distinct from previous models." It was a good start for the new company.

## W & X

- Red-Chestnut Overo Pinto Polo Pony, 883-18; May 2006, 25 made; VG: $23… EX: $30… NM: $38.

Each one of these models looks different due to random, overo pinto markings. They would all have in common: a medium, reddish chestnut body; black hooves; pink muzzle (assuming that all of the models have the same, white face marking); and darker brown knees, hocks, pasterns, mane, and tail.

The white markings on this model's body look like they were created by dabbing off the reddish paint with a sponge or other small implement. The white areas appear to be the unpainted, white plastic.

Overo pinto markings are often relatively smallish and ragged white spots on a dark field; they appear to hang from the sides of the horse and do not usually cross the back. This model portrays that type of pinto.

This model was announced in May 2006, and arrived in June 2006. The price was $30. It was a run of 25, and 17 were still left in March 2007.

**Y**
Of the16 Polo Ponies, nine have white face markings, including: the Bay Tobiano Pinto (883-13), Red Roan (883-15), Red-Chestnut Overo Pinto (883-18), and Dapple Palomino with green wraps (883-19).

**Z**
More production Polo Ponies with white face markings are: the Black Appaloosa (883-03), Liver-Chestnut Pinto (883-04), Pale Palomino (883-05), Buckskin Pinto (883-08), and Black Pinto (883-09).

**AA**
**Spotted Polo Ponies** – White markings typically cross the top line of tobiano pintos, but not overo pintos. *From left* are: the Black Appaloosa (883-03), Liver-Chestnut Overo Pinto (883-04), Buckskin Overo Pinto (883-08), Black Overo Pinto (883-09), and Bay Tobiano Pinto, 883-13.

**BB**
**Palomino Polo Ponies** – "Volant," the 883-19 Dappled Palomino with Green Wraps *(left)* is darker than the Pale Palomino, 883-05.

## POLO PONIES: 9" SERIES – PRODUCTION MODELS
### LEG AND TAIL WRAPS ( BANDAGES ) ARE WHITE IF COLOR IS NOT STATED

| Color & Name | Item # | Run Type | Debut Date | Qty. |
|---|---|---|---|---|
| Copper-Bay Appaloosa with Pale Gold Wraps, "Chart the Course" | 883-01 | Club model, 2001 | Feb. 2001 | 270 |
| Flecked Grey with Turquoise Wraps | 883-02 | r | June 2001 | 50 |
| Black Appaloosa with Pale Gold Wraps | 883-03 | r | July 2001 | 75 |
| Liver-Chestnut Overo Pinto | 883-04 | r | Nov. 2001 | 50 |
| Cinnamon Dapple, "Sugarplum" | 883-06 | Holiday horse, 2001 | Nov. 2001 | 150 |
| Dapple Grey with White Wraps | 883-07 | r | Feb. 2002 | 50 |
| Buckskin Overo Pinto, "Jammer Time" | 883-08 | Ticketed Jamboree special, 2002 | Aug. 2002 | 100 |
| Pale Palomino | 883-05 | r | Sept. 2002 | 35 |
| Black Overo Pinto, "Black Tie" | 883-09 | Show special model, 2003 | June 2003 | 78 |
| Metallic Copper with White Mane and Tail, "Cinn" | 883-17 | Club model, breed series horse, 2004 | March 2004 | 85 |
| Metallic Gold Bay, "Heart of Gold" | 883-16 | Holiday horse, 2004 | Nov. 2004 | 85 |
| Dun / Grulla | 883-10 | r | Dec. 2004 | 39 |
| Bay Roan | 883-15 | r | Dec. 2004 | 44 |
| Bay Tobiano Pinto | 883-13 | r | Dec. 2004 | 44 |
| Red-Chestnut Overo Pinto | 883-18 | r | May 2006 | 25 |
| Palomino (Dappled) with Green Wraps, "Volant," 2nd in "Noble Horse Series" | 883-19 | Sold to winners of mail-in drawing | Jan. 2007 | 15 |

**Polo Ponies** – Hartland made more colors of 9" Polo Ponies from 2001-2007 than any other horse shape. The 16 production colors include five pintos, two appaloosas, three dappled colors, and six others. Nine colors were regular runs, and seven were special runs, including club models for 2001 and 2004, holiday models for 2001 and 2004, and the show special model for 2003. Four of the Polo Ponies were editions of 100 or more, and "Chart the Course," with 270, was the second largest Hartland horse edition since 2000. (That doesn't count the two horse-and-rider sets.)

**Chukker, Anyone?** The first model horse from Hartland Collectibles, L.L.C., was a Polo Pony: the bay blanket appaloosa, "Chart the Course," in February 2001. He (or she) led a string of 16 production Polo Ponies, enough for four teams in play at once.

Other colors for 2001 were a dark-flecked grey, black blanket appaloosa, liver-chestnut overo pinto, and a pearled "cinnamon dapple." A dapple grey, buckskin pinto, and pale palomino left the barn in 2002, followed by a black overo pinto in 2003.

Five colors cantered onto the field in 2004: metallic copper, metallic gold bay, dun / grulla, bay roan, and bay pinto. A red-chestnut overo pinto joined them in 2006. Finally, a dappled palomino with green wraps chased the ball through the goal posts in January 2007.

# CHAPTER 8
# HORSES WITH REINS
## 9" RIDER SERIES

*Roy Rogers and Dale Evans rode back into town in new outfits. Trigger and Buttermilk had new tack. Bullet was with them. Four saddled horses waited at the hitching rail.* Six new western sets released by Hartland 2000 included horses of two shapes: Semi-Rearing (8.25" tall x 9" long) and Chubby Standing / Walking (8.25" tall).

**A**
This "Dale Evans, Buttermilk, and Bullet," 9" rider set #802, was released in 2005.

**B**
This "Roy Rogers and Trigger," 9" rider series set #806, appeared in 2005.

**C**
Roy Rogers is all smiles on his third release since the 1950s. He and Dale Evans have eye color painted in for the first time, and their boots have never been so elaborately painted.

**D**
The 2005 model of Bullet, Roy Rogers' German Shepherd, has brown shadings. (In the 1950s-1960s, Bullet's shadings were dark gray to black.)

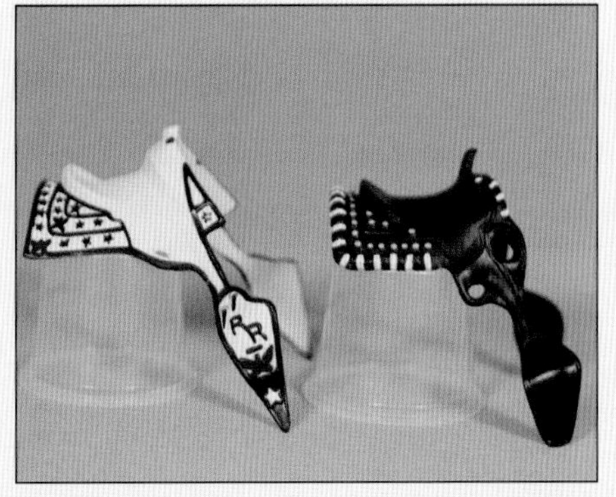

**E**
Roy Roger's saddle is white and very dark blue with red accents; it's the version with rounded fenders. Dale Evans' saddle is black with white stripes; the mold is the cowboy saddle with rifle hole.

**F**
Small parts of the 2005 Roy Rogers and Dale Evans rider sets are: his white hat and her reddish hat, and three identical, silver-painted pistols. Two pistols are his.

## TWO ENTIRE RIDER SETS

Hartland's 9" rider series returned in a small, but nice, way after 2000, proving that there is a trail back after riding off into the sunset. In December 2005, Hartland's Roy Rogers and Dale Evans horse-and-rider sets were released for the third time since the 1950s. Their attire was painted differently from all previous versions, and for the first time, their eye color and other facial features were painted in.

## B & H

- Roy Rogers and Trigger, #806; December 2005; an edition of 2,500 sets.

**Roy Rogers.** Roy is all smiles at his comeback. His teeth are painted white. His eyes have an aqua blue iris and black pupil. His eyebrows are painted brown and his lips are darker flesh color than his face.

His mold is the square-shirt-yoke version, which dates to 1958 (*Hartland Horsemen*, page 151.) His shirt is red and white; his boots are red, white, dark blue, and black. He came with a white hat and two silver pistols.

The saddle is the Roy Rogers saddle version with rounded fenders and molded-in cinch holes, which was last used in 1992-1994. A fancy parade saddle, the color scheme is white with red eagles, and dark blue (almost black) stars and trim.

**Trigger.** Trigger is a 9" series Semi-Rearing horse with Mane Up and Plain Tail, in an attractive, opaque and metallic palomino color. His coat is golden yellow, apparently with reflective, metallic gold flecks in the finish. (Possibly, the only Hartland palominos that are more attractive are the translucent, glossy palominos from the 1950s and early 1960s.)

Trigger's mane, tail, and left hind sock are painted white. He has a wide, white painted-on blaze that turns to pale peach on the muzzle, and there's a pale spot on the lower lip.

The hooves do not match the usual hoof colors since 2000. The left hind hoof is a different beige-pink than usual, and the other three hooves are light brown (more specifically, dark rosy taupe).

The bridle is painted white with the brow band outlined in dark blue. A silver dot is painted at each end of the brow band, and the "snaffle bit" rings are painted silver. The painted-on breast collar has a red eagle with six blue stars on each side of the eagle.

The reins are a white plastic lace that is tied closed. The reins on these newly released horses are very flexible because the plastic lace is so new and quite thin.

## A & H

- Dale Evans, Buttermilk, and Bullet, #802; December 2005; an edition of 2,500 sets.

**Dale Evans.** This likeness of Dale Evans has aqua blue irises with black pupils, brown eyebrows, and lips painted red. The figurine shows a combination of brush painting and airbrush painting. The mold, which has bangs, dates to 1957; it is the second mold used to portray Dale (*Hartland Horsemen*, page 151).

Dale wears dark blue culottes, a dark blue tie, and a white blouse with red collar and cuffs. Her boots are red, white, and dark blue with a butterfly motif. She came with a red hat and one silver pistol.

Her saddle is the cowboy saddle with rifle hole, painted black with white stripes around the edge; that is, on the corona (saddle) pad, and small silver dots on the back saddle skirts.

**Buttermilk.** Dale's horse, Buttermilk, is the Chubby mold with molded-on bridle headstall and normal tail for that mold. The color is pale buckskin or creamy dun. This horse's body is pale cream with a smooth, but not highly glossy, finish. The mane, tail, and feet were painted very dark brown, which is not a hoof color seen on other Hartland horses since 2000. Buttermilk also has brown shading on the muzzle and nostrils, above the eyes, and inside the ears. The ear tips are dark brown – on both on the front and the back of the ears!

**Bullet.** The German Shepherd, Bullet, has a cream-colored body except for dark brown shadings on the back, sides, face, muzzle, and upper tail. He has a black muzzle and light gray feet except for brown toenails. His eyes have a round, black pupil inside a round brown iris, and eye whites showing in both the front and back corners of his eyes.

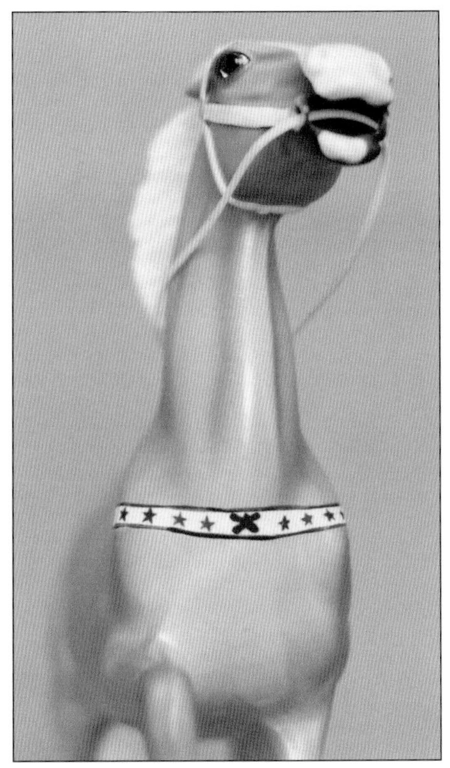

**G**
The white on Trigger's face extends to his muzzle and lower lip. The painted-on breast collar is white with dark blue edges and a red eagle flanked by dark blue stars.

**H**
Roy Rogers and Dale Evans ride Trigger and Buttermilk, 2005. This Trigger is the first Hartland production model with a breast collar, instead of a martingale neck loop.

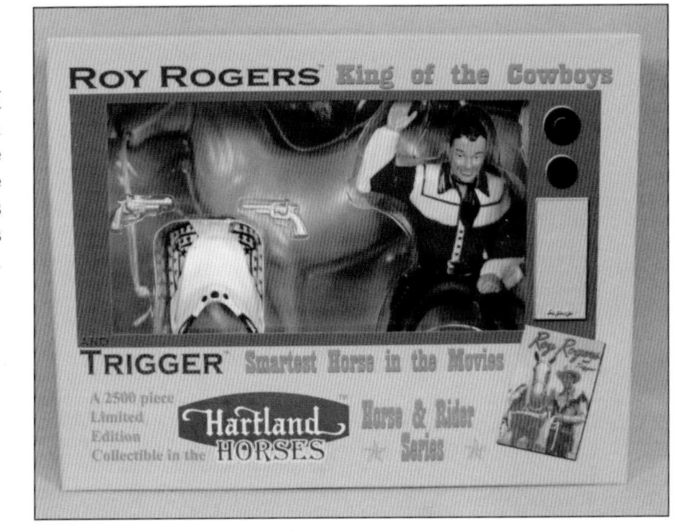


**I**
Roy Rogers waves from his box. His and Dale Evans' 2005 boxes are styled as TV screens with the figurines visible.

**J**
Dale Evans waits with Buttermilk and Bullet in their 2005 box, which calls her, "Queen of the West."

**K**
One panel of the box for Roy Rogers and Dale Evans illustrates their saddle. Each set is called, "A classic collectible plastic figurine set."

**L**
The art on the 2005, Roy Rogers and Dale Evans rider set boxes was also used on the 1950s-1960s Hartland boxes. One difference is that the 2005 box tells "The Hartland Story."

**Boxes and Stickers.** Roy and Dale each came in a decorative box with a clear-view front resembling a TV set. The art on the back of the boxes is the same painting of Roy and Dale on horseback that appeared on the 1950s-early 1960s boxes.

The boxes read, "A classic collectible plastic figurine set by Hartland Horses™," and "Made in China." There's a cautionary note: "This collectible contains small parts and is not recommended for children under 12." There is trademark information: "Dale Evans, Buttermilk, Roy Rogers, and Trigger Trademark owned by ROY ROGERS and used by Horse-Power Graphics, Inc. under authorization." Below the "Hartland Horses" name is, "Hartland Trademark owned by Hartland Collectibles L.L.C. used by permission," and the company's web address, www.hartlandcollectibles.com. A side panel promotes the Roy Rogers and Dale Evans Museum & Happy Trails Theatre in Branson, Missouri. (The museum closed in 2009.)

A sticker on the boxes advised using a square of wax paper "or some other type of protective barrier" to separate the rider, saddle, and horse "to prevent paint transfer." (That advice would also apply to any Hartland saddle released since 2000.)

The Roy, Dale, Trigger, Buttermilk, and Bullet figurines each have a sticker on the underside that reads, "© 2003 Horse-Power Graphics, Inc." The stickers are silver-colored and one-half inch wide.

Hartland said that the designs of the post-2000 Roy Rogers and Dale Evans sets were approved by Roy "Dusty" Rogers, Jr.

**Museum Auction.** The Roy Rogers and Dale Evans Museum, which had been operated by the Rogers family, closed at the end of 2009, after 42 years. On July 14-15, 2010, most of the museum contents were sold by Christie's auction house in New York. An auction lot that included the 2005 Hartland Roy Rogers and Dale Evans sets and other Roy Rogers branded merchandise sold for $3,250. A lot picturing the 2005 Hartland "Buttermilk," six Breyer horses, and other toys, sold for $1,875. (Auction results are from www.christies.com.)

**A Difficult Trail.** It was a rocky road back for Roy and Dale. The company took deposits of $10 each for Roy and Dale in January 2004, and said it expected delivery the following month. Then, the sets were delayed due to a legal problem in which another company that was licensed to produce Roy Rogers items complained that Hartland was also licensed. The deposits were returned (in cash) in June 2004. Finally, the sets were available in December 2005. All's well that ends well.

**Prices and Values.** The Roy and Dale sets were each an edition of 2,500 pieces. In December 2005, each set was $50 from Hartland Horses, and combined shipping was $10.95. In December 2010, each set was $45 plus shipping from Horse-Power Graphics on eBay; both sets together were $80 with free shipping. In January 2012, sets were still available at those prices.

In 2010, collector Sandra Truitt informed me of an alternate source for the 2005 Roy and Dale sets: RFD-TV, a cable TV network in Nebraska. Its online store sells the sets, which it calls "old-time action figures," for $35 each, plus shipping. Those prices were in effect in June 2011. In January 2012, the sets were $30 each.

For "Roy Rogers and Trigger," 9" rider series set #806, suggested values are:

Complete set with all parts and box: VG: $54... EX: $72... NM: $90.

Individual components:

Roy, the rider, VG: $12... EX: $16... NM: $20.
Roy's saddle, VG: $11... EX: $14... NM: $18.
Roy's hat, VG: $3... EX: $4... NM: $5.
Roy's two pistols, NM: $2 each.
Roy's box, NM: $3.
Trigger, VG: $24... EX: 32... NM: $40.

For "Dale Evans, Buttermilk, and Bullet," 9" rider set #802 , suggested values are:

Complete set with all parts and box: VG: $48... EX: $64... NM: $80.

Individual components:

Dale, the rider, VG: $11... EX: $14... NM: $18.
Dale's saddle, VG: $6... EX: $8... NM: $10.
Dale's hat, VG: $3... EX: $4... NM: $5.
Dale's pistol, NM: $2.
Dale's box, NM: $3.
Bullet, the dog, VG: $9... EX: $12... NM: $15.
Buttermilk, VG: $16... EX: 22... NM: $27.

## Extra Horses

The four "extra" horses, each with a plastic rein and removable saddle, are:

### O

- Pale Buckskin 9" Semi-Rearing Horse, "Buckeye," 850-B; 2004 Western Club Member Model, March 2004, 75 made; VG: $24... EX: $32... NM: $40.

"Buckeye" is a pale, golden-beige dun or buckskin. His body is banana yellow (the inside of a banana, not the skin) with pale, rosy shadings and no dorsal stripe. He has brown shading on the knees, hocks, eye area, nostrils, and lips, and a dark brown mane, tail, and hooves, including the hoof bottoms.

His bridle is painted solid black with the bit area (molded in a snaffle-ring shape) painted silver. He came with a dark brown saddle, rust brown reins tied in place, and dark blue reins as an alternative or for use as a cinch strap.

"Buckeye" went to the 2004 Hartland "western" club members for $50. His picture appeared on the 2004 lapel pin along with the Polo Pony that was the 2004 "horse" club model. (Double membership was $80.) He was first shipped in March 2004.

"Buckeye" is edition-numbered in the groin area. Example: #055.

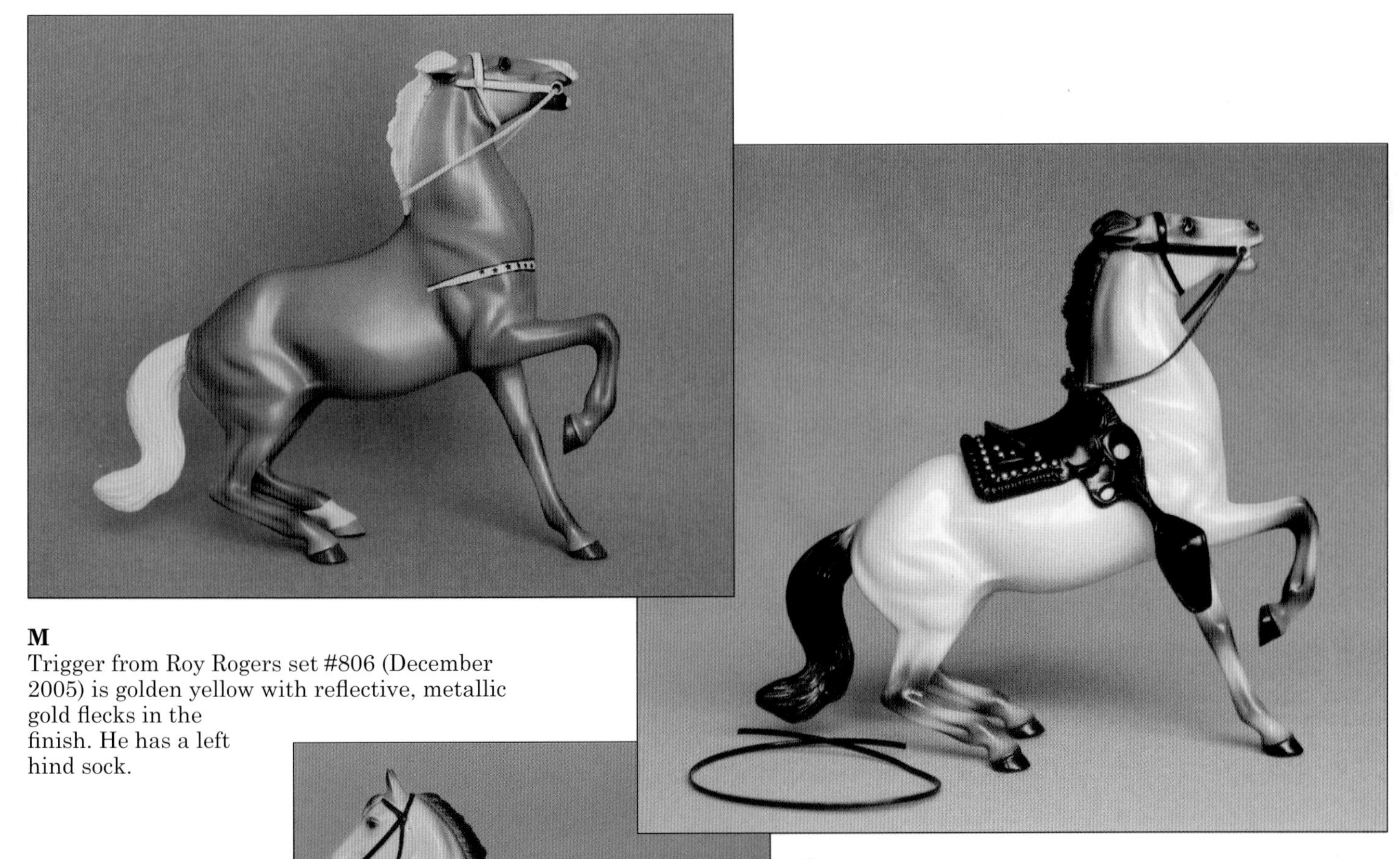

**M**
Trigger from Roy Rogers set #806 (December 2005) is golden yellow with reflective, metallic gold flecks in the finish. He has a left hind sock.

**N**
Buttermilk from Dale Evans set #802 (December 2005) is the only horse from Hartland 2000 with very dark brown hooves. The ear tips are brown, too.

**O**
The Pale Buckskin 9" Semi-Rearing Horse, "Buckeye," 2004 Club Model (rider series option), 850-B, came with two pairs of reins.

## P

- Chocolate Bay 9" Semi-Rearing Horse, "Cocoa," 850-CB; December 2007, 20 made; VG: $30... EX: $40... NM: $50.

This model has an even, chocolate brown body color with no shadings; a dark muzzle, and black points. The bridle is solid black with the bit area painted silver. It has a black saddle and black reins.

"Cocoa" came out after the company's final newsletter issue, and the company did not publish a catalog number for it. So, I took the mold number, 850, and added "CB" for "Chocolate Bay."

"Cocoa" was announced late in December 2007, and was sold on a first-come, first-served basis, for $45 plus $9.95 shipping. He was a numbered edition of 20. An example of the numbering is: 19/20.

Since the release of "Cocoa," more than four years have passed without the release of any new Hartland horse or western items.

**P**
Chocolate Bay 9" Semi-Rearing Horse, "Cocoa," 850-CB.

## Q

- Coffee Dun 9" Chubby Horse, "Java Joe," 820-CD, April 2005, 56 made; VG: $24... EX: $32... NM: $40.

In April 2005, Hartland came out with another coffee dun; this time, on the Chubby Standing/ Walking horse. There are two mold variations of the Chubby dating to the 1950s. The type used since 2000 has a molded-on bridle headstall.

In the final *Hartland Courier*, the company listed this model's number as "821-CD," but I'm sure that was a typographical error, one of several in that list of models. In the *Hartland Courier*, Spring 2001, the company said the general Chubby mold number was 820, which makes sense. The number 821 belonged to a 1950s-1960s rider set, Tom Jeffords (*Hartland Horsemen*, p. 127). I think "Java Joe" was meant to be number 820-CD, and so I'm calling it that.

(To add to the confusion, in the model list in the final *Hartland Courier*, the company also swapped the names of the two coffee dun horses! I checked the list against all of the other company-produced sources. The other sources were all consistent, and the opposite of what was published in the company's final list.)

**Q**
Coffee Dun 9" Chubby Horse, "Java Joe," 820-CD.

**R**
Coffee Dun 9" Semi-Rearing Horse, "Cuppa Joe," 2005 Club Model, 850-CD.

"Java Joe" has black points and hooves and a black bridle with a silver-painted bit area, and a silver dot on the cheekpiece, near the bit, and on the browband, at the cheekpiece. The hoof bottoms are painted black, too. The saddle and reins are black.

On all (both types of) Chubby horses, the tail is molded separately from the body, and then inserted. Now, some of the models painted and sold since 2000 were actually molded in the 1990s. The previous owner of Hartland, Steven Manufacturing Company, had made Chubby horses and also Standing / Walking horses with inserted tails. When Hartland Collectibles, L.L.C., bought Hartland from Steven in 2000, it acquired old stock, including unpainted and unassembled models and unattached tails.

In 2005, at least some of the "Java Joe" Chubby horses, including the one in photo C, received the wrong tail by mistake. It has the tail for the Standing/Walking horse (with wavy tail).

It's a different silhouette and, in effect, a new "mold variation" for the Chubby. The tail hangs more angled back than it should because the Chubby's tail hole is set higher than the Stander / Walker's. Dale Evans' horse "Buttermilk," in photo N, has the correct Chubby tail.

The coffee dun Chubby was $35, and 56 were made. The "CD" suffix on the model number evidently stands for "Coffee Dun." This model was not sequence-numbered on the underside. It had a name, but was a regular run.

## R & T

- Coffee Dun 9" Semi-Rearing Horse, "Cuppa Joe," 850-CD; 2005 Club Model, March 2005, 100 made; VG: $24... EX: $32... NM: $40.

"Cuppa Joe" was the 2005 club model, came out in March 2005, and was limited to a run of 100. The $45 club membership included this horse and a silver rifle with Hartland logo.

(In 2005, "Cuppa Joe" was the only club model. In 2004 and in 2006, there was a choice of two club models: a breed-series horse or a western series item – or you could pay more and get both. In 2001-2003, a breed-series horse was the only club model.)

"Cuppa Joe" has black points and a body the color of coffee with cream, a soft brown that is darker and richer than taupe. Since the points are black, and there's no dorsal stripe, this model could be called "buckskin," rather than "dun" for those reasons, but "buckskin" often conjures up a golden body color whereas "dun" is often assigned to drabber shades.

In any event, Hartland Plastics, Inc., made two models in this color in the early 1960s. Both were for rider Gil Favor and are hard to find. In the early 1980s, collector Jaci Bowman called the color "coffee dun." I've used that term in Hartland books since 1983, and in 2005, Hartland followed suit, calling it, "'Cuppa Joe,' Coffee Dun Semi-Rearing Horse with saddle." (I coined the term, "Semi-Rearing," in the 1983 book, too. Hartland had not named most of its rider-series horse molds.)

One of the early 1960s horses in coffee dun was a 9" series Semi-Rearing horse, but with mane down. That shape is found on some test models released since 2000, but "Cuppa Joe" and the other Semi-Rearing production models are of the Mane-Up, Plain Tail type. (The third type of Semi-Rearing horse has Mane Up and Fancy Tail. A coffee dun in that mold shape was made by Paola Groeber's Hartland Collectables as a five-piece test run in 1986. It has hind socks.)

The molded-on bridle on "Cuppa Joe" is painted black; the bit area (snaffle shaped) is painted silver, and there's a silver dot on the cheekpiece, near the bit, and one on the browband at the cheekpiece. The reins and saddle are black.

The model has black on the muzzle, around the eyes, and inside the mouth. The black color on the legs was sprayed on, but on the mane and tail, the black color was brush-painted by hand.

The Coffee Dun Semi-Rearing horse is hand-numbered on the underside. Example: #045.

**Saddles for the Extra Horses.** The saddles that came with the four, unmanned horses are all of the same shape: the cowboy saddle with rifle hole. (See photo S.) These saddles have a cinch hole on each side and, on the right side only, a larger hole representing a rifle boot. (The rifle that came with a number of the 1950s-1960s rider sets could be inserted, held in place, there.)

Although all four saddles are of the same shape, their color and finish were not the same.

"Buckeye," the pale buckskin horse, has a dark brown saddle; it was painted over clear-colored plastic. The clear plastic shows through where a flake or two of paint came off.

The saddles of the two coffee dun horses, "Cuppa Joe" and "Java Joe," are the same. They are painted black with a smooth, but not glossy, finish. The saddle for "Cocoa," the chocolate bay, is different: it is painted glossy black.

All four saddles have dark undersides, with nothing hand-written on them. The saddles of the two coffee duns are painted black on the underside, with white plastic visible in scratched areas. The underside of the saddle for "Cocoa" looks like black paint over dark plastic.

**S**
The saddles of the four independent horses are, *from left*: glossy dark brown for "Buckeye"; non-glossy black for the two coffee dun horses; and glossy black for "Cocoa," the chocolate bay.

**T**
Coffee Dun 9" Semi-Rearing Horse, "Cuppa Joe," 850-CD – *left side.*

**Values.** The values for the four extra horses include the saddles. If missing the saddle, subtract $4 from the horse's value.

**Different from Extra Saddles.** In contrast, the black and light brown rifle-holed cowboy saddles that were sold without a horse in this century are white on the underside and have, "Hartland 2002" written in gold-colored ink. See photos in the following chapter.

**Horses with Reins.** The reins through the mouth of the 9" rider series horses give them extra "play" value even though they are fine enough to be kept on a pedestal. Before 2000, earlier Hartland companies made over 70 horse-and-rider sets, in 4 sizes (scales), using 11 basic horse molds (more than 20 molds, counting mane, tail, and other shape variations). The rider series through 1999 is thoroughly explained in *Hartland Horsemen*. The recent models are a nice addition to the vast body of earlier work.

| HORSES WITH REINS ( 9" RIDER SERIES ) - PRODUCTION MODELS<br>FOUR HORSE-AND-SADDLE SETS AND TWO HORSE-AND-RIDER SETS | | | | |
|---|---|---|---|---|
| **Color & Name** | **Item #** | **Run Type** | **Debut Date** | **Qty.** |
| Pale Buckskin Semi-Rearing Horse, "Buckeye" | 850-B | 2004 club model, rider series option | March 2004 | 75 |
| Coffee Dun Semi-Rearing Horse, "Cuppa Joe" | 850-CD | Club model, 2005 | March 2005 | 100 |
| Coffee Dun Chubby Horse, "Java Joe" | 820-CD | r | April 2005 | 56 |
| Roy Rogers and Trigger horse-and-rider set | 806 | r | Dec. 2005 | 2,500 |
| Dale Evans, horse Buttermilk, and Bullet, their German Shepherd dog | 802 | r | Dec. 2005 | 2,500 |
| Chocolate Bay Semi-Rearing Horse, "Cocoa" | 850-CB | r | Dec. 2007 | 20 |

**Rider Horses** – In 2004-2007, Hartland released four horses with reins and saddles and two complete horse-and-rider sets, all from the 9" rider series. The Roy Rogers and Dale Evans rider sets, editions of 2,500, were the only new "mass produced" plastic horse items from Hartland 2000.

# Chapter 9
# Rider Accessories and Old Stock

In the 1950s-1990s, molded plastic accessories were part of the horse-and-rider sets. In the new century, Hartland 2000 issued some 9" series accessories independently of rider sets or horses, intending them as replacement parts, and also sold some old 1990s stock "as is," which was, in many instances, unpainted.

**A**
The Black *(left)* and Brown Cowboy Saddles with Rifle Hole were sold individually. They differ from saddles that Hartland 2000 sold with a horse.

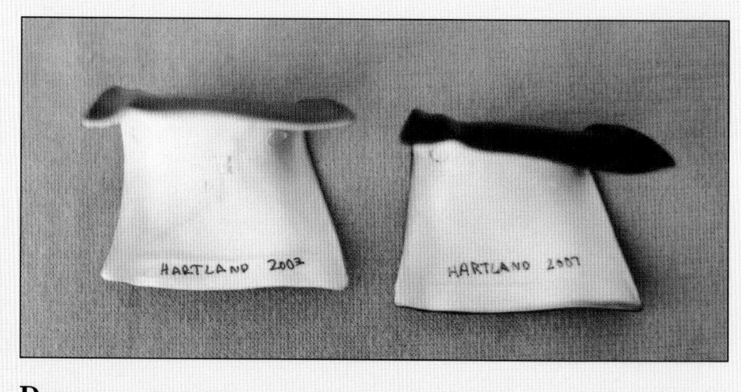


**D**
The underside of the Brown Indian Saddle *(left)* reads, "Hartland 2003." The Black Indian Saddle *(right)* is marked, "Hartland 2007."

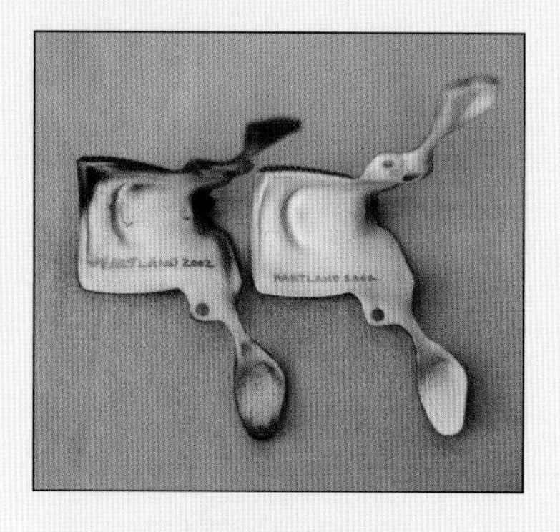

**B**
The underside of each Black or Brown Rifle-Holed Cowboy Saddle sold alone is labeled "Hartland 2002."

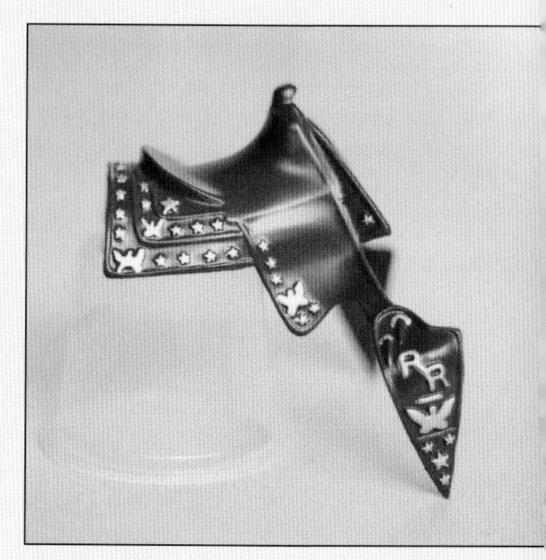

**E**
The Roy Rogers Saddle with Squared Fender, sold alone (starting in April 2005) is painted dark blue with silver trim.

**C**
The Brown Indian Saddle *(right)*, was sold alone. The Black Indian Saddle *(left)* was part of a six-piece set.

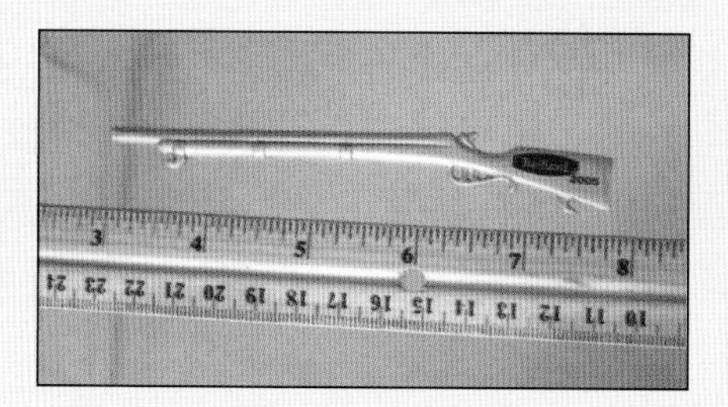

**F**
The Rifle with Hartland Logo came with the 2005 club membership. It is painted silver and reads, "Hartland 2005."

**Saddles Sold Separately.** Four saddles were sold individually since 2000. Their quantities were not announced, but they were plentiful. On all saddles since 2000, tiny details like silver dots or stars were painted freehand, with a brush. From previous Hartland companies, the details on 9" series saddles were airbrushed using masks. The saddles fit well on the 9" rider-series Hartland horses from the 1980s on, but look a little large on their 1950s-1960s counterparts.

## A & B

- Black Cowboy Saddle w/ Rifle Hole.
- Brown Cowboy Saddle w/ Rifle Hole.

The Black and Brown Cowboy Saddles with Rifle Hole, June 2002, were $25 each. They were on sale for $18.75 from November 15-December 20, 2002, and then the price went to $20.

They have a smooth, but not glossy finish, and "Hartland 2002" handwritten in gold on the underside. (The rifle-holed saddles sold with a horse since 2000 are not marked like that.) They are called "cowboy saddles with rifle boot" to distinguish them from other styles of western saddles used with rider sets since the 1950s.

In May 2004, 91 were for sale in an eBay auction by the company (eBay I.D.: hartland_horses). That total included both colors: black and brown. They were apparently sold out by December 2010. The value for these particular, hand-dated saddles is: VG: $12... EX: $16... NM: $20 each.

## C & D

- Brown Indian Saddle (sold alone).
- Black Indian Saddle (part of an accessory set).

The "Native American tan saddle blanket" has "HARTLAND 2003" written on the underside. It was $15 in May 2003. In 2009, it was $10; by December 2010, they were apparently sold out.

The Black Indian saddle was part of Indian Accessory set #2, "Black Indian Accessories with Saddle," and is described with that set.

If Hartland 2000 had sold an Indian saddle with a horse or a complete rider set, it's value would have been about $3 or $4. However, for the Brown Indian Saddle, I suggest: VG: $4... EX: $6... NM: $8; MIP:$10.

## E

- Roy Rogers Saddle w/ Squared Fender

The dark blue Roy Rogers saddle with squared (pointy) fenders came out in April 2005 for $25. (The fenders are the saddle flap on each side that rest beneath and behind the rider's upper leg and knee. The term applies to western saddles. Another mold of Roy Rogers saddle, which was used most recently for the 2005 Roy Rogers and Trigger set, has rounded fenders.)

This saddle with square fenders is dark blue with silver detailing: stars, eagles, and "RR" initials. In February 2009, it was available for $15. On May 20, 2011, the price was higher ($20), and in January 2012, more than 10 were available. For this particular saddle, I suggest: VG: $12... EX: $16... NM: $20.

**Besides Saddles.** More accessories sold apart from a complete rider set included rifles, more Indian accessories, and key chains made from hats.

## F

- Rifle w/ Hartland Logo; Club Rifle for 2005

This silver-painted rifle, almost 5" long, came with the 2005 club membership. The vinyl label on it reads, "Hartland 2005." The quantity was probably 100 since 100 club models were made. Values for this particular rifle are: VG: $3... EX: $4... NM: $5.

## G – upper hat

- Hat Key Chain with Tag

When Hartland took $10 deposits on Roy Rogers and Dale Evans sets in January 2004, it promised a hat key chain with the first 100 orders. The rider sets were delayed, but the key chain was sent in June 2004 with the refunds. The hat is unpainted, white plastic with "© Roy Rogers used with permission" printed on the tag. It was made by punching a hole in a Dale Evans hat. Values for hats with a key chain hole are: VG: $3... EX: $4... NM: $5.

**G**
Dale Evans hats were issued as key chains with paper tag in 2004, and with "Hartland Horses 2006" on the crown in 2006.

## G – lower hat

- Hat Key Chain with Hartland Logo; Club Key Chain for 2006

The key chain hat marked with a dark blue Hartland logo, "Hartland Horses 2006," came with either club model selection for 2006. (Those who selected both models received two key chains.) The logo was printed on a piece of clear vinyl applied to the unpainted hat, which was a Dale Evans hat with a hole punched in it. The quantity made was between about 100 and 150. They were issued starting in January 2006, and sent to those who joined the club any time that year. Values for hats with a key chain hole are: VG: $3... EX: $4... NM: $5.

## H & M

- Black Indian Accessories with Saddle (6 pc), August 2007; Indian accessory set #2.

This set for $30 in August 2007 included six pieces: the Indian saddle, headdress, rifle, short bow, long bow, and shield. The company called it: Native American Accessory Set #2.

The color scheme is black and white, with a black-and-silver rifle. The headdress is the low-rise headdress, Brave Eagle's. The saddle is also shown in photos C & D.

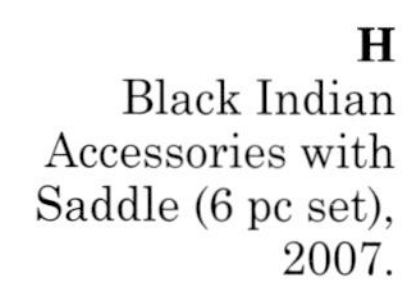

**H**
Black Indian Accessories with Saddle (6 pc set), 2007.

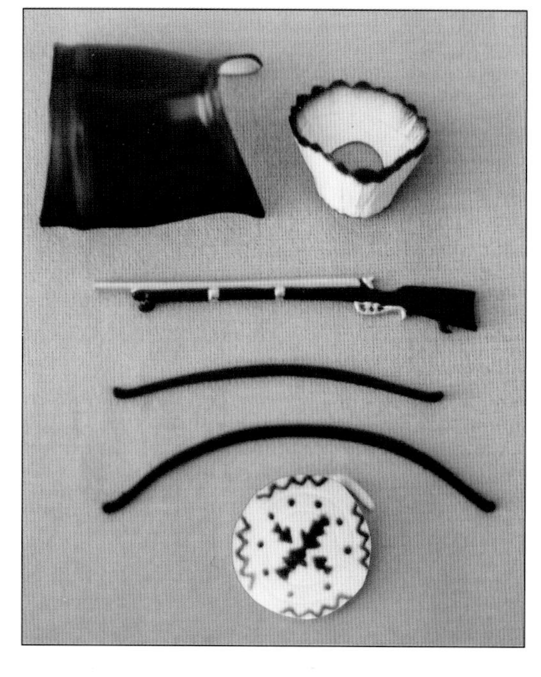

**I**
Black Indian Accessories with Spear (6 pc set), 2003.

The quantity was not stated; they were plentiful. In February 2009, the sets were $25. In May 2011, more than 10 sets were available for $25. Six were for sale in January 2012.

In the painted Indian accessory sets in general, the small parts like a bow or knife don't have as much eye appeal as larger parts like a shield, headdress, or spear, but were more easily lost years ago. For that reason, and to keep things simple, each piece in the sets is considered to have equal value. The average value of each piece, new in package, in this particular Indian accessory set was $5. Per piece: VG: $2... EX: $3... NM: $4...Mint:$4.50; set, MIP: $30.

# I

- Black Indian Accessories with Spear (6 pc), May 2003; Indian accessory set #1.

This black and white set in May 2003 was $30. By February 2009, the price was $25. The quantity was not announced; the supply was plentiful. Native American Accessory Set #1 includes six pieces: a long knife, short knife, and tomahawk painted black and silver; and a feather, headdress, and spear in black with areas of white, unpainted plastic. The headdress shape is Chief Thunderbird's. In January 2012, more than 10 sets were available for $25.

The average value of each piece, new in package, was $5. Per piece: VG: $2... EX: $3... NM: $4; Mint:$4.50; set, MIP: $30.

# J

- Brown Indian Accessories (6 pc), Indian accessory set #3; August 2007, 50 made
- Brown-and-Silver Rifle; August 2007; 50 made.

The six-piece, brown-and-white Indian accessory set was $30 in August 2007. The company called it: Native American Accessory Set #3. It was an edition of 50. Those who ordered both this set and the black Indian accessories (set #2) in 2007 were promised a bonus item: a rifle painted brown and silver. The rifle actually arrived heat-sealed in its own compartment in the clear plastic package with the brown accessories. So, the brown set could be considered seven pieces. Also, I wonder: If anyone bought only the brown set and not the black set, did they get the rifle anyway? Either way, the number of brown rifles produced should be 50.

Apart from the rifle, the other six pieces are: a long knife, short knife, and tomahawk painted brown and silver; and a headdress, feather, and spear painted brown, with the white areas being the unpainted plastic. The headdress shape was Chief Thunderbird's.

The three, painted Indian headdresses sold after 2000 are compared in photos K & L.

By February 2009, the price was $25. In May 2011, more than 10 of the six-piece Brown Indian Accessories sets were available, for $25. Five sets were still available in January 2012, without the rifle.

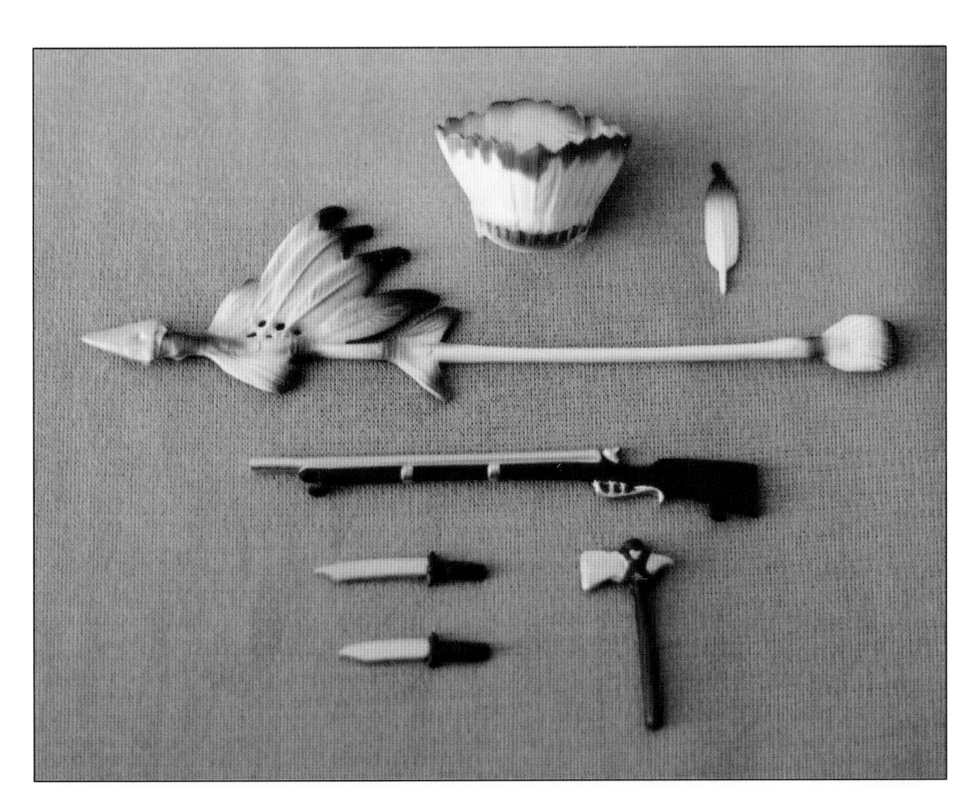

**J**
Brown Indian Accessories, six pieces plus a brown-and-silver rifle, 2007.

On December 1, 2010, the Brown-and-Silver Rifle was for sale by itself, for $6. Ten rifles were available; the auction stated that those were the last remaining. Values for rifle alone: VG: $3... EX: $4... NM: $5... Mint: $5.50.

For the set of six (or seven) pieces, the values per piece are: VG: $3... EX: 4... NM: $5; Mint: $5.50. The seven pieces (including rifle), MIP: $40.

## Vintage and Values

**Molded Earlier.** Probably all of the accessories sold by Hartland 2000 were molded in the 1990s by Steven Manufacturing Company, the previous owner of Hartland. The accessory items painted in colors never seen before were painted by Hartland 2000. From mid-2008 on, the only original source of new or old Hartland stock from the Hartland company was Horse-Power Graphics, Inc., selling them chiefly on eBay.

**K**
Hartland 2000 issued three painted headdresses in Indian accessory sets. The high-rise headdresses in brown and black *(left and center)* were Chief Thunderbird's; the low-rise headdress (*right*) was Brave Eagle's.

**Values.** Values for these parts are a difficult call. When the Hartland rider sets were new in the 1950s and 1960s, the saddle was a small part of the set's value. Because so many saddles were lost or broken, the old saddles with stirrups intact often sell for $20 or more. When Hartland 2000 sold saddles separately, the price was $15-$20 or more. Ironically, the Hartland 2000 horse-with-saddle sets were $35-$45, so the saddle was, at most, maybe $5 of the value. The entire Roy Rogers and Dale Evans rider sets from Hartland 2000 were $50 each, so, again, the saddle might be worth $5. (It would make little difference that the Roy and Dale sets were made overseas, but the four horse-and-saddle sets were made in the United States.) There's no doubt about it: The stand-alone saddles from Hartland 2000 were overpriced. That is in spite of the fact that they do not fit well on the sets that need them (1950s-1960s), and the sets they do fit (1980s-1990s) are few and not widely collected.

The same is true of the other painted accessory parts Hartland 2000 sold without a complete set: They were overpriced, and there isn't much use for them. However, the lone saddles and other parts could be considered collector items in their own right – and if you wait 20 years, they might shrink enough to fit the old horses and riders a little better. So, the values given for them will usually be at or close to their latest selling price, which in most cases is from 2009-2011.

**L**
Brave Eagle's headdress *(right)* is both lower and longer (more horizontal) than the Chief Thunderbird headdresses *(left and center).*

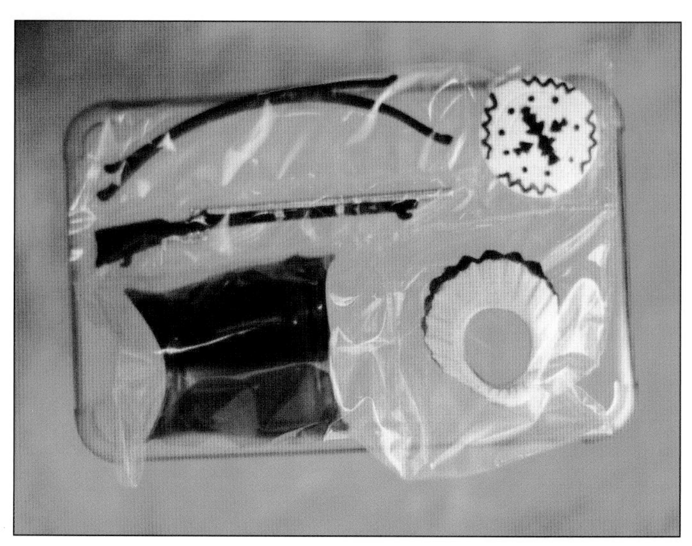

**M**
The accessory sets came in a partitioned, clear plastic bag with the parts heat-sealed into separate compartments. This is the Black Indian Accessories with Saddle 6-piece set.

| RIDER SET ACCESSORIES: 9" SERIES - PRODUCTION MODELS | | | | |
|---|---|---|---|---|
| **Description** | **Item #** | **Run Type** | **Debut Date** | **Qty.** |
| Black Cowboy Saddle with Rifle Hole | -- | r | June 2002 | many |
| Brown Cowboy Saddle with Rifle Hole | -- | r | June 2002 | many |
| Brown Indian Saddle | -- | r | May 2003 | many |
| Black Indian Accessories w/ Spear (6 pc) | Indian acces. set #1 | r | May 2003 | many |
| Dale Evans Hat Key Chain w/ Tag (No logo on hat) | -- | Went w/ first 100 Roy/Dale deposits | June 2004 | 100 |
| Rifle w/ Hartland Logo and "2005" | -- | Went w/ club model, 2005 | March 2005 | 100 |
| Roy Rogers Saddle w/ Squared Fender (dark blue saddle) | -- | r | April 2005 | many |
| Dale Evans Hat Key Chain w/ Hart. Logo | -- | Went w/ club model, 2006 | Jan. 2006 | 100 to 150 |
| Chief Thunderbird Tree (Unpainted Indian & accessory parts from 1990s) | -- | Club "model" 2006, rider series option | Jan. 2006 | many |
| Black Indian Accessories w/ Saddle (6 pc) | Indian acces. set #2 | r | Aug. 2007 | many |
| Brown Indian Accessories (6 pc) | Indian acces. set #3 | r | Aug. 2007 | 50 sets |
| Rifle: brown & silver | -- | Enclosed w/ Ind. acces. set #3 | Aug. 2007 | 50 |

**Equipment –** Most of the newly-painted saddles and other 9" rider-set items sold without a horse in 2002-2007 were unlimited editions and, apparently, plentiful. At least some, if not all, of them had been molded in the 1990s, but since they were painted after 2000, they count as new production. The unpainted, 1990s Indian "tree" is included in the table because it was a "club special."

## Old Hartland Stock Sold in the New Century

In addition to newly painted rider accessories, some old (1990s) Steven Hartland stock of western series items was sold in the new century. Some items were painted, but most were unpainted white. Where it's noted that items went up for sale in November 2005, they were first advertised in a postal mailing from Horse-Power Graphics, and the quantities available were not announced. Where it's noted that they were still available in 2009 or later, it means that Horse-Power Graphics was selling them on eBay. Also, Hartland sold some old stock of breed-series horses starting in late 2000.

**N**
The Chief Thunderbird Tree molded in 1994 was a club "model" for 2006, and was also sold after 2006.

**O**
Unpainted, 1990s rider series items sold after 2000 included, *from left:* Gen. Lee with hat, sword, cinch, and flag; Dale Evans with hat; and Gen. Washington with hat, sword, and flag.

## N

- Chief Thunderbird "Tree" (Unpainted Parts)

The Western club membership "model" for 2006 was Chief Thunderbird as molded, before the parts were detached, assembled, and painted. The Indian was molded in front and back halves. The other parts on this mold were (clockwise, from top) a bow, tomahawk, spear, rifle, shield, headdress, saddle blanket, and knife. It was old stock molded in the 1990s by Steven Manufacturing Company.

The 2006 Western Club membership was $55, and also included a hat key chain with Hartland logo on it. (Those who bought both club models in 2006 – this tree and a 9" Tennessee Walker – paid a total of $80.)

The quantity of trees was not announced. In February 2009, the parts, broken off the tree, were available for $25, and intact trees were $40. By May 2011, the "kits" of loose pieces were $20, and intact trees were $35. In June 2011, more than 10 of each were still available.

Values: Intact tree: $35; the group of 10 loose pieces: $20.

**P**
The Chief Thunderbird and red saddle painted in 1994 and the Four-Saddle Tree were available after 2000. The white saddles are: the Roy Rogers saddle with squared fender *(upper left),* the Mountie saddle *(lower right),* and two Indian blankets. Hartland molded its saddles with the stirrups sticking out sideways, and then reshaped them.

**Unpainted, White Riders**

## O

- The Gen. Lee rider, hat, sword, and cinch strap (all unpainted, white) and polyester Confederate flag went up for sale in November 2005 for $35. In May 2011, the set was $15 and more than 10 were left. In January 2012, you could also buy just the hat and cinch for $5.
- The Gen. Washington rider, sword, and hat, all unpainted, and with a 13-star polyester flag, was $25 in November 2005. The rider was also untrimmed; that is, had rough seams. The 13-star "Betsy Ross" flag by itself was $5.

In July 2009, the unpainted Washington and Lee rider-and-accessory sets were sold together for $40. In May 2011, the Washington set was $15, and the flag by itself was $2.50. More than 10 sets and 10 flags were still available.

- The Dale Evans rider and hat in unpainted white appeared on eBay in December 2010. I believe the pieces were molded in the 1990s. The price from Horse-Power Graphics was $15. In May 2011, more than 10 were left. In January 2012, another option was Dale's hat for $5.

Values in condition as molded by the factory: Lee and Washington sets (with as-new flags), $25 each; Dale Evans set (no flag), $15.

## P – center

- 1994 Chief Thunderbird Rider and Saddle (Painted)

The 1994 Chief Thunderbird rider and saddle blanket (red), both painted, with possibly flaking paint, were $25. They went up for sale in November 2005. By March 2011, the price was $15, and in June 2011, seven were left. Collectors could read in my book *Hartland Horsemen* (1999) that fewer than 100 complete sets (with horse) were made, so this color scheme of the Indian rider was rare. He has war paint on the sides of his face. Value: $15-$25, depending on the condition of the paint.

**Unpainted, White Saddles**

## P – left and right

- A Four-Saddle Tree in Unpainted White.

Two Indian Saddles ("blankets"), one Canadian Mountie saddle, and one Roy Rogers saddle with pointy (squared) fender were together on the same mold. In 2006, the company awarded one "tree" with those four saddles to (Mr.) Lonnie Chapman, whose name was drawn from among two people who submitted a picture or write-up about their collection after the company's Winter 2006 newsletter invited readers to do so. Then, in December 2010, the Four-Saddle Tree was $30 on eBay; in May 2011, 10 trees remained.

The unpainted saddles were also sold individually, as follows:

- The Roy Rogers Saddle with pointy (squared) fender was $10 in unpainted white in December 2010, and some were still available in May 2011.
- The Chief Thunderbird saddle blanket (unpainted, white plastic) was $8 in November 2005. Later, it was $5. In December 2010, it was $2; more than 10 were still left in May 2011.
- A Canadian Mountie saddle, unpainted and untrimmed was $15. (Untrimmed means that the seams were not smoothed.) By December 2010, the price was down to $5, and more than 10 were left in May 2011.

The suggested value for the intact, Four-Saddle Tree is $30. For separate, unpainted white saddles, values are: Roy Rogers saddle, $10; Mountie saddle, $8; and Indian saddle, $4.

## Q

- Unpainted Wyatt Earp Gunfighter

Starting in July 2002, the Wyatt Earp standing gunfighter, unpainted and with arms unattached, in white plastic, from 1994, was available for $25. From November 15-December 20, 2002, it was $18.75. In January 2003, the price went to $20. It was still available in November 2004, but was sold out by 2009.

Values: still sealed in package, $25; in factory condition, but no package, $22.

More 1990s Hartland Items sold since 2000 are:

- Two George Washington saddles, one unpainted, and one painted, which could have flaking paint, were $20 for the pair in November 2005. In May 2011, the price was $10 and more than 10 were still left. Value: painted or unpainted, $5 each.
- In July 2009, a group of six saddles was $40. The group consisted of: two unpainted Mountie saddles, two unpainted Indian blankets, an unpainted Washington saddle, and a painted Washington saddle.
- Dale Evans' brown saddle from the 1990s set was for sale in December 2010 for $24.99. The mold is the cowboy saddle with rifle hole. It does not match any saddles painted after 2000.
- The 1994 Wyatt Earp Rider sets, painted and new-in-box, were available from Hartland starting in late 2000. On eBay, they were $55 in 2009, and apparently sold out in 2010.
- Trigger, fully painted in palomino color by Steven Manufacturing in the early 1990s, went on sale in November 2005 without his rider or saddle for $25. He was described as semi-rearing with mane down, and sold "as is." On an order form that arrived in February 2007, he was $10. This version of Trigger is pictured in the Introduction.

**More Rider Series Models.** Painted, complete sets of western models from the 1990s are pictured in *Hartland Horsemen* (Schiffer, 1999). This chapter and Chapter 8, "Horses with Reins – 9" Rider series," describe the western-series items that were production items and/or were abundantly for sale from 2000 on. Test (unique) western models are in Chapter 17. There were also some rare (but not unique) western series items; they are in chapter Chapter 18, "Other Unusual Models."

**1990s Breed Series Horses.** Hartland 2000 also sold some leftover, 1990s Steven Hartland breed-series horses new-in-package starting in late 2000. That is described in the "History" chapter, and the horses are illustrated in *Hartland Horses and Dogs* (Schiffer, 2000). Also, some paint sample 1980s-1990s horses were sold in 2010. (See Chapter 18.)

**Still Available After 2007.** Since 2008, the leftover stock from the 1990s and from 2000-2007, has been sold on eBay by Horse-Power Graphics, Inc., whose owner was horse director for Hartland 2000. (The eBay seller name is: horse_power_graphics_inc.) Among them, in June 2011, was the new-in-box, 1990s 9" series Grazing Arabian Mare in rusty grey. Nine were available for $19.99. In January 2012, there were eight left.

**Q**
Unpainted, Wyatt Earp Standing Gunfighters molded in the 1990s were sold after 2000, unassembled. The arms can move at the shoulder like doll arms. The gunfighters can't ride.

## CHAPTER 10

# TINYMITE HORSES

### SIX BREEDS 3" LONG

*In the distance, approaching horses made a moving quilt of ebony, copper, and gray; browns, reds, and yellows.* The Tinymite horses are six breeds 2.25" to 2.75" tall and 3" long: the Arabian, Tennessee Walker, Morgan (with head tucked down), Thoroughbred, Quarter Horse, and Belgian Draft Horse. Six new sets and additional colors were woven into the tapestry of Tinymite history after 2000.

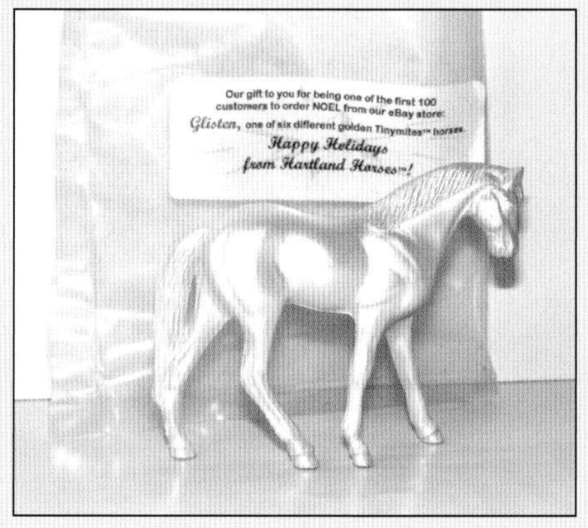


**A**
This is a Tinymite Thoroughbred in pale gold. A pale gold Tinymite in one of six breeds went with each purchase of "Noel," a 9" holiday model.

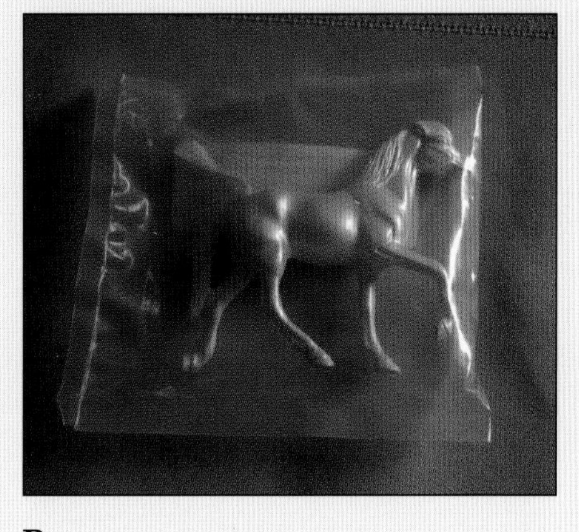

**B**
All six Tinymite breeds in pale gold had the same name, "Glisten." This is the Tennessee Walker.

**C**
The label on the package with this Pale Gold Tinymite Draft Horse has a different word arrangement than the one in photo A.

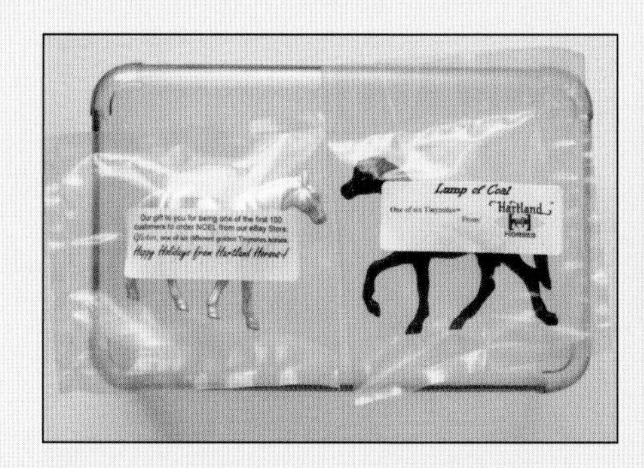


**D**
"Glisten" and "Lump of Coal" came packaged individually. A "Lump of Coal" (black Tinymite) came with "Sugarplum," a 9" holiday horse.

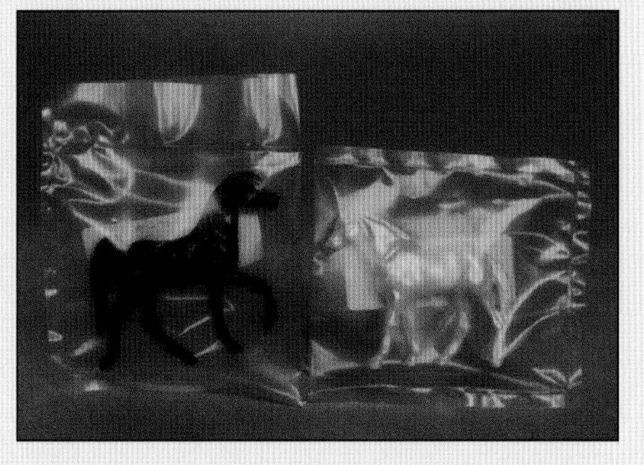

**E**
The "Lump of Coal" and "Glisten" here and in photo D are the Tennessee Walker *(left)* and Quarter Horse. (Not shown: pale gold Arabian and Morgan.)

**F**
Black Tinymites – *Top left*: Belgian, Thoroughbred and Morgan; *bottom row:* Arabian, Tennessee Walker, and Quarter Horse.

The Tinymites came in two color types: "solid" colors, meaning that they are one color from head to foot with no details painted in, and "non-solid" colors, which are the ones in realistic colors, with painted details.

## A-E

- Pale Gold Tinymites, "Glisten" (six breeds); TM-6G; went singly with 2001 eBay Holiday Special, November 2001; 20 made of each breed; VG: $12... EX: $16... NM: $20; MIP:$22.

A pale golden, pearled Tinymite, "Glisten," was supposed to go with the first 100 orders of "Noel," the 9" 2001 eBay Holiday model, but "Noel" turned out to be a run of only 85. Yet, the company said it made 120 pale gold Tinymites: 20 of each of the six Tinymite breeds. So, where did the others go? Well, one of each breed (a group of six) was sold on eBay with the first "Noel" model, so that accounts for five more of the pale gold Tinymites, for a total of 90. The extra 30 pieces were never advertised.

When ordering "Noel," there was no choice of which breed of pale gold Tinymite came with him. Read more under "Noel" in Chapter 3 and Holiday Models in Chapter11. *Photos, courtesy of: Elaine Boardway (A), Robyn Porter (B), and Denise Hauck (C).*

## D, E, & F

- Black Tinymites, "Lump of Coal" (six breeds); TM-6B; went singly w/ 2001 Holiday Horse, November 2001; 50 made of each breed; VG: $3... EX: $4... NM: $5; MIP: $7.

Each order of "Sugarplum," the 2001 Holiday horse (a 9" Polo Pony) was accompanied by a "Lump of Coal" – a glossy black Tinymite. There was no choice of breed. "Sugarplum" could be ordered from November 2001 until Jan. 31, 2002, for $36. After the 2001 holiday season, the six breeds of black Tinymites went up for sale individually, for $3 apiece, in February 2002. Fifty were made of each breed. The total run of 300 black Tinymites was sold out in 2002. Their color looks molded in, rather than painted on. Read more under "Sugarplum" in Chapter 7 and Holiday Models in Chapter 11.

## G

- Blue Roan Tinymite Draft Horse, TM-1D; October 2003; 35 made; VG: $12... EX: $16... NM: $20.

The Blue Roan Tinymite Draft Horse was a surprise special run at Jamboree 2003; 35 were made. (An earlier report by the company said 32.) It has a grayish body with black points and head. Its tail ribbon is painted white. It sold out at the event. The price was never published; a collector recalled it as $20. *Photo G, courtesy of Pat Noble.*

## H

- TinyMite Draft Horse Set (5 pc), TM-5D; October 2003; 50 sets made; Per Horse: VG: $5... EX: $7... NM: $9; MIP set: $50.

The five-piece TinyMite Draft Horse set was the ticketed, Hartland special run for Jamboree 2003 (October 24-26, in Pomona, California). It had to be pre-ordered from Horse-Power Graphics, Inc., for $32.50 and picked up at the event. The TinyMite Draft Horse Set – notice that the company spelled it still as one word, but with a capital "M" – was an edition of 50.

The five draft horses are:

Dapple Grey – with black lower legs, black hooves, and tail ribbon painted light blue.
Palomino – with dark gray shading on the muzzle, four white socks, peach hooves, and red ribbon; the company called it: blond sorrel.
Chestnut Roan – with black hooves and medium green ribbon.
Bay – with black hooves, four white socks, and yellow ribbon.
Black – with peach hooves, four white socks, and dark blue ribbon.

This set was a numbered edition. The number was written on a card inside the models' package.

# I

- "The Tiny Five" Tinymites (5 pc, five breeds); TM-51; October 2004; 50 made; Per Horse: VG: $5... EX: $7... NM: $9; MIP set: $50.

The Arabian in this Tinymite set is dapple grey. This set was the pre-ticketed special run at Jamboree 2004 (October 8-10, in Pomona, California) The set had to be advance ordered, for $32.50 from the Jamboree, and picked up at the event. Only 50 sets were made.

The breeds and colors are:

Quarter Horse – Bright Bay with hind socks and peach hooves under the socks; otherwise, the points and hooves are black.
Thoroughbred – Chestnut Roan with black hooves.
Arabian – Dapple Grey with four white socks; black mane, tail, knees, hocks, muzzle, and hooves.
Morgan – Buckskin with no white.
Tennessee Walker – Pale palomino with four white stockings; pink muzzle and hooves.

"The Tiny Five" was a numbered edition. The number was written in gold ink on the ticket, which was returned to the person picking up the model, with "void" stamped on the back of it.

**G**
The Blue Roan Tinymite Draft Horse was sold individually, not as part of a set.

**H**
The "TinyMite Draft Horse Set" (5 pc) consists of, *top:* Dapple Grey and Palomino; *bottom:* Chestnut Roan, Bay, and Black draft horses.

**I**
"The Tiny Five" Tinymites are, *top:* a Bay Quarter Horse and Chestnut Roan Thoroughbred; *bottom:* Dapple Grey Arabian, Buckskin Morgan, and Palomino Tennessee Walker.

**J**
"The Might-y Five" Tinymites are, *top:* a Blue Roan Quarter Horse, Chestnut Arabian, and Palomino Thoroughbred; *bottom:* Bay Tennessee Walker, and Dapple Grey Morgan.

**K**
Copper Tinymite 5-pc set, "Frankincense," the 2003 Holiday Horse Set consists of, *top:* Arabian and Morgan; *bottom:* Quarter Horse, Thoroughbred, and Tennessee Walker.

**L**
The "Christmas Chocolates" 5-pc Tinymite sets vary in their mix of light brown and dark brown models. Here, the Morgan *(top)* and Tennessee Walker are dark brown.

## J & P

- "The Might-y Five" Tinymites (5 pc, five breeds), TM-52; October 2004; 50 sets made; Per Horse: VG: $5... EX: $7... NM: $9; MIP set: $50.

This is the Tinymite group with a chestnut Arabian. It was a surprise, special run of 50 sets at Jamboree 2004 (October 8-10, at Pomona, California). The price was $32.50 plus 8.25% California sales tax.

The breeds and colors are:

Quarter Horse – Blue Roan with black points.
Arabian – Flaxen, Liver Chestnut.
Thoroughbred – Pale Palomino with both pink and brown shading on the muzzle.
Tennessee Walker – Bright bay with left hind sock with peach hoof below it and three black hooves.
Morgan – Dapple Grey with four white socks, and a black mane, tail, knees, hocks, and hooves.

"The Might-y Five" was not a numbered edition.

## K

- Copper Tinymites, "Frankencense" (5 pc set), TM-5C; December 2003, Holiday Horse Set, 30 made; Per Horse: VG: $4... EX: $6... NM: $8; MIP set: $45.

The copper-colored Tinymites were "Frankincense," the 2003 Holiday Horse Set, consisting of five breeds: all but the Draft Horse. The color is so even all over, including the hoof bottoms, that it might be molded in.

This set was available from December 1, 2003, through January 31, 2004, for $32.50 plus $6.95 shipping. The company set a limit of 250 sets, but the eventual edition size was 30. The set was edition-numbered on a card enclosed with it.

## L, M, & N

- Light Brown and Dark Brown Tinymites, TM-55; "Christmas Chocolates," (5 pc set); Per Horse: VG: $4... EX: $6... NM: $8; with box: add $2.

December 2005, Holiday Horse Set, 30 made

This 2005 Holiday Horse set, "Christmas Chocolates," was available from late December 2006 until January 31, 2006. The first sets arrived in January. The price was $32.50. The final edition size was 30.

This five-piece set included one of each Tinymite breed except the Belgian. In the sets I've seen, two breeds were dark brown, to resemble dark chocolate, and three were light brown, like milk chocolate. Which breeds were light or dark varied from set to set. The horses were nested in gold tissue paper in a clear-topped box, tied with a cord. One piece of actual, chocolate candy was enclosed.

The light brown and dark brown Tinymites were definitely painted over white plastic. The edition number for each set was written in gold ink on a business-sized card enclosed in the box. *Photo M, courtesy of Robyn Porter.*

**M**
The "Christmas Chocolates" set came in a box with gold foil wrap and a clear plastic cover. This example has a dark brown Arabian and Thoroughbred.

**N**
This third example of the "Christmas Chocolates" Tinymite set has a dark brown Quarter Horse and Arabian. A piece of real chocolate wrapped in foil *(center)* came with it.

**O**
Five Tinymite sets came with identity cards about 3.5" x 2". The set's edition number is hand-written in gold ink on three of them. The card *at bottom left* was an advance-purchase ticket.

**P**
All of the Tinymite sets except "Christmas Chocolates" were packaged in clear plastic, with the identity card and each horse heat-sealed in separate compartments. This set, with chestnut Arabian, is "The Might-y Five."

**Tinymites for Jamboree.** Four special-run Tinymite items were connected with the Jamboree: two in 2003 and two in 2004. Here's their story. Each of those two years, there was one Tinymite set that was a ticketed, special run that had to be pre-ordered. It was sold by the Jamboree itself (Horse-Power Graphics, Inc.). Also in each year, there was a surprise, special run Tinymite item that was sold by the Hartland company, from a booth at the Jamboree.

In 2003, tickets for the "TinyMite Draft Horse Set" special run set for Jamboree went on sale in September 2003 for the October 24-26 event. The company newsletter said that the tickets for the 50 sets sold out in 24 hours, weeks before the Jamboree.

After that news, the remaining draft horse Tinymites were painted blue roan, the newsletter said, and Hartland sold them from a booth at the Jamboree. The Blue Roan Draft Tinymites were sold out during the event.

In 2004, essentially the same thing happened. The Jamboree (Horse-Power Graphics) sold advance tickets for the special run of 50 sets of "The Tiny Five." The tickets were sold out at least three weeks before the October 8-10 Jamboree; the company newsletter reported that they were sold out in one day.

So, then, Hartland prepared a second set of the five breeds, "The Might-y Five" and sold them as a surprise special run, from the Hartland booth at Jamboree. The surprise set sold out by noon on the first day of the show, according to the company newsletter.

So, in 2003, there was a draft horse set and a single (blue roan) draft horse, and in 2004, there were two sets with five, different light-horse breeds. When you bought the ticketed sets, you were buying them from Horse-Power Graphics, a Hartland distributor; the surprise items at Jamboree were sold by the Hartland company, itself. Since the owner of Horse-Power Graphics was the director of Hartland's horse line, it was a very tightly knit situation.

**Tinymite Eyes and Other Details.** A contrasting color was painted on the tail ribbon of the Tinymite draft horses. The shape of the ribbon was part of the mold created by Hartland Plastics sculptor Roger Williams in the 1960s, but it was never painted in a contrasting color until 2003.

On the Blue Roan Draft Horse Tinymite sold individually, the eyes are black with a white back corner, and the hooves are black. On the three realistically painted sets – draft horses and light horses – the hooves were either peach or black. There was usually darker shading on the muzzle (unless the whole head was black, and except on some palominos). Here's a summary of how the eyes were painted, and the muzzle color on the palominos:

- "TinyMite Draft Horse Set" – In this set from 2003, the eyes are black, with or without white or peach painted in the back corner. The palomino has black on the muzzle.
- "The Tiny Five" – In this set from 2004, the eyes were painted black, with the back corner of each eye painted white, or in some instances, peach. The palomino's muzzle has pink shading.
- "The Might-y Five" – In this surprise, special run set from 2004, the eyes were painted the same way as in the "The Tiny Five" set. The palomino in this set has both pink and pale brown muzzle shading.

Company literature described "The Tiny Five" as [cellulose] acetate plastic, the better plastic.

**Packaging and Paperwork.** Most of the Tinymite sets were packaged in a heat-sealed, partitioned, clear plastic bag measuring about 8" x 10". The exception was the "Christmas Chocolates" set, which came in a clear-topped box. The black and golden Tinymites sold individually at Christmas 2001 came sealed in a single-horse-sized plastic bag.

The cards that came with Tinymite sets are about 3.5" x 2". Two of the cards list the model colors, and three cards include an edition number hand-written in gold ink. The other sets were not numbered editions. Each card pictured in photo O is described below:

- The card from "TinyMite Draft Horse Set" (October 2003) lists the colors of the five Draft Horses. The sets in this edition were not numbered.
- The card from the "Frankincense" set of five copper Tinymites (the 2003 Holiday Horse Set), is thinner paper than the others. It has a dark green background, so the edition number is written in the white box at the bottom of the card. The set illustrated was "009" out of 30.
- The card from "The Tiny Five" Set, the pre-ticketed set of five Tinymites sold at Jamboree 2004, lists the model colors in addition to picturing the models.
- The card for "The Might-y Five" Set pictures the five Tinymites, but does not list their colors. This set was a surprise special run at Jamboree 2004. It was not a numbered edition.
- The ticket for "The Tiny Five" Set was bought in advance, presented at the 2004 Jamboree, stamped "void" on the back, and returned to the buyer. The sequence (edition) number was written by hand in the white box on the ticket. The gold ink color is hard to read on the example illustrated, but it is "013."
- The card with the "Christmas Chocolates" Set includes the set number written in gold ink on the dark green part of the card. The card was enclosed in the box with the five "chocolate" Tinymites, which were the 2005 Holiday Horse Set.

**The Breeds.** The black and pale gold Tinymites produced in 2001 were the first new ones since the 1970s. The new company applied a more generic label to some of the breeds. It called the Belgian, the "draft horse"; the Quarter Horse, the "stock horse"; the Thoroughbred, the "warmblood"; and the Morgan, the "prancer." It still referred to the Arabian and Tennessee Walker by their original, 1960s appellations. This book uses the original breed names.

**Catalog Numbers.** The company did not assign catalog numbers to the Tinymites, so I created codes for them. Each code starts with "TM" for Tinymite, followed by a hyphen and a code for the set's size and color (or breed).

They are:

TM-6B = black color, found on six breeds ("Lump of Coal")
TM-6G = pale golden color, found on six breeds ("Glisten")
TM-5D = five draft-horse set, each horse a different color
TM-1D = one draft horse (blue roan)
TM-5C = five breeds, all in copper color
TM-51 = the first set consisting of five different breeds that are each in a different color ("The Tiny Five," 2004 ticketed special run)
TM-52 = the second set consisting of five different breeds that are each in a different color ("The Might-y Five," 2004 surprise special run)

TM-55 = five breeds in a mix of chocolate colors (either light brown or dark brown)

To specify an individual model, it's necessary to add the breed or color or both after the code. Examples:

TM-6B – Thoroughbred (from the all-black set)
TM-5D – Bay (from the all-draft-horse set)
TM-51 – Dapple Grey Arabian (from "The Tiny Five" mixed-breed and mixed-color set)

| TINYMITE HORSES: SIX BREEDS 3" LONG – PRODUCTION MODELS | | | | |
|---|---|---|---|---|
| **Description** | **Code** | **Run Type** | **Debut Date** | **Qty.** |
| Black, "Lump of Coal" – 6 breeds sold singly w/ "Sugarplum" 9" Polo Pony (6 breeds for $3 each in March 2002) | TM-6B | Went w/ 2001 Holiday horse | Nov. 2001 | 50 of each breed |
| Pale Gold, "Glisten" – 6 breeds sold singly with "Noel" 9" TWH | TM-6G | Went w/ 2001 eBay Holiday horse | Nov. 2001 | 20 of each breed |
| "TinyMite Draft Horse Set"– 5 different colors | TM-5D | Ticketed Jamboree special, 2003 | Oct. 2003 | 50 sets |
| Blue Roan Draft Horse (one horse) | TM-1D | Jamboree 2003 surprise special | Oct. 2003 | 35 |
| Copper Set, "Frankincense," 5 breeds | TM-5C | Holiday Horse Set, 2003 | Dec. 2003 | 30 sets |
| "The Tiny Five" – 5 breeds, each a different color (w/ dapple grey Arab) | TM-51 | Ticketed Jamboree special, 2004 | Oct. 2004 | 50 sets |
| "The Might-y Five" – 5 breeds, each a different color (with chestnut Arab) | TM-52 | Jamboree 2004 surprise special | Oct. 2004 | 50 sets |
| "Christmas Chocolates" – 5 breeds: some light brown, some dark brown | TM-55 | Holiday Horse Set, 2005 | Dec. 2005 | 30 sets |

**Tinymites –** The eight Tinymite production items released from 2001-2005 were all special runs. They included five, five-piece sets; one Tinymite sold alone; and two Tinymites that were paired with a larger model.

**Pocket Size.** In the world of plastic horses, small size never goes out of style.

# CHAPTER 11
# DETAILS AND EXTRAS

*There were arched necks, dappled sides, flared nostrils, and soulful eyes. Everywhere, coats gleamed in the sun.* The production models featured pearled finishes, dapples, face markings, and other painted details. This chapter includes extra views and additional production items, such as resin horses connected with Hartland. It describes the eyes on the 9" and 11" series models, highlights the special editions, and summarizes the production models in tables.

**Dapples.** Most of the dappled horses by Hartland 2000 are "resist" dappled. Collector Eleanor Harvey explained that the technique begins with spraying a light mist of oil on the model. Then, when paint is sprayed on, the oil droplets resist the paint since it is based on a different solvent. Instead of sticking to the oil, the paint "pools around the globular edges of the oily spots, creating the halo or 'resist' dapple effect as the paint dries," she said. "Once the paint has dried, the oil is wiped off, and the model is washed." Then, what some painters will do is spray on a thin top coat of varnish or sealant to protect the paint, she said.

Eleanor said that resist dappling is not new, but dates back to 1960s Breyer "Decorator" horses. "It's the most efficient method for a factory setting, where the goal is speed and relative consistency." No two "resist" dappled models will look exactly alike, but then again, none of the standard dappling methods result in identical horses, Eleanor said.

A more time- and labor-intensive dappling technique was apparently used on the two horses in Hartland 2000's "Noble Horse Series." In the "fish-scales" technique, "the painter airbrushes tiny arcs of a darker color in a light, overlapping pattern on the model," Eleanor said. "These can be closed circles or scale-like "C" shapes, depending on the artist."

If each dapple was painted individually, no wonder the "Noble Horse Series" was priced higher!

**Face Markings.** Face markings can enhance a model's appearance, and many of the 7", 9", and 11" production models by Hartland 2000 have them. The Tinymites do not. Face markings usually look better when they are painted on, rather than created by merely masking the area to keep it unpainted. The technique of painting on the star, stripe, blaze, or snip was used on all but two of the Hartland production models with face markings released in 2001-2007.

The exceptions were two models whose face markings are in conjunction with elaborate pinto markings: the Buckskin Overo Pinto and the Red-Chestnut Overo Pinto. The white areas on the face and body of those models are the glossy, unpainted white plastic.

**Hoof Bottoms.** Another area of thorough attention to detail was hoof bottoms. The bottoms on raised feet were always painted. Additionally, the soles of standing feet were painted on some Tinymites and on three of the five 9" series models released in December 2004: the Bay Pinto 9" Five-Gaiter, Buckskin 9" Tennessee Walker, and Bay Tobiano Pinto Polo Pony.

**Dapples --** The Black (Dappled and Pearled) 7" Tennessee Walker Family Mare *(top)* and Stallion *(bottom)* from set 684-02, July 2001, exhibit abundant dapples.

The foal from the Bay (with Pearled undercoat) 7" Tennessee Walker Family, set 684-04, looks iridescent in this photo.

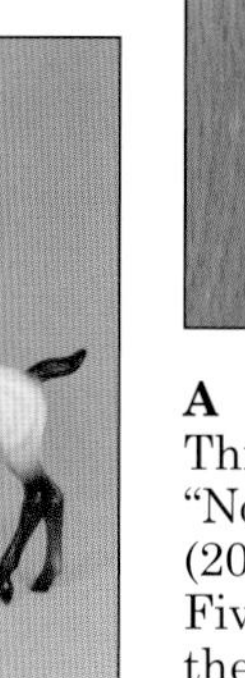

The Bay Roan 7" Tennessee Walker mare and foal from set 684-06 have reddish-brown on the head as bay ("red") roan horses do.

**A**
Three 9" holiday models wore neck tags: "Noel," a dapple grey Tennessee Walker (2001); "Snow Angel," a dapple grey 9" Five-Gaiter (2002); and "Heart of Gold," the metallic gold bay 9" Polo Pony (2004).

**B**
Horses on the "club" lapel pins were: 2001 – "Chart the Course," the bay appaloosa Polo Pony; 2002 – "Wave the Banner," the chestnut pinto Five-Gaiter; 2003 – "Steadfast Heart," the dapple grey Arabian; and for 2004, two models: "Buckeye," the pale buckskin semi-rearing horse, and "Cinn," the metallic copper Polo Pony.

**C**
A subscription to the company's newsletter, *Hartland Courier,* was part of its annual "club" membership. This is the Spring 2005 issue with the dark blue 11" Arabian, "Hart of Blues."

**D**
The "Walter" mold, a trotting stock horse gelding in resin, was to join the Hartland line, but it was issued by its sculptor, Sheryl Leisure, instead, in colors including palomino *(shown).*

**E**
This cap from Hartland 2000 has a portrait of the 11" Quarter Horse embroidered on the front, and the Hartland logo on the back.

**F**
"Seabiscuit," a resin 3.5" tall and 5.75" long, was mass produced in China, and released in July 2003, shortly before *Seabiscuit* opened in theatres.

**G**
The "Seabiscuit" box has the "Hartland Collectibles" name and story on it. The model was sold by numerous eBay sellers, but not through the Hartland web site or mailings.

**H**
This logo key ring from Hartland 2000 includes a leather part 4.25" long and 1.5" wide.

## Wearable or Usable Items

In September 2001, the company website offered an embroidered, denim long-sleeve shirt with the Hartland logo for $25. In December 2010, Horse-Power Graphics was selling it on eBay for $25. Other miscellaneous production items included:

### A

- Neck Tags

Three of the 9" holiday models wore neck tags about 1.5" square. NM tags: $2 each.

### B

- Lapel Pins

For four years, the Hartland "club" membership included a cloisonné lapel pin picturing the special run club model or models. Pins, NM: $5 each.

### C

- Newsletter

The company's annual club membership included a subscription to its quarterly publication, *Hartland Courier.* There were 22 issues, usually six or eight pages each, and 146 pages in all, including four pages in full color. EX-NM: 20 cents per page or $30 for all.

### E

- Cap with Visor

A design of the head and neck of the 11" Quarter Horse is embroidered in white on the front of this cocoa brown ball cap, and "Hartland" is on the back. It was produced by American Needle. The cap was $15 on the Hartland company website in September 2001. It was free with a horse or western order of $50 or more between mid-November and December 15, 2005. In June 2011, nine caps were still available for $10 on eBay. AV-EX: $3-$8; New, never worn: $10.

### H

- Leather Key Ring

Another promotional item was a leather key ring with the Hartland name on it. The holder is leather, 4.25" long and 1.5" wide, according to Robyn Porter. Robyn thought hers was a door prize donated by Hartland to the Northwest Congress model horse show she and Melissa Clegg held in May 2002. Elaine Boardway said that Hartland gave her a couple of key rings to use as door prizes at her model horse show in May 2001. *Photo H, courtesy of Robyn Porter*. AV-EX: $3-$8; New: $10.

## Extra Horses

Counted as "extras" are resin models that Hartland advertised, but did not produce, and a resin model sold under the Hartland name, but that was not really part of the Hartland horse line.

### D

- "Walter" and Other Resin Models

Three new molds sculpted by Hartland's horse director, Sheryl Leisure, were advertised by Hartland in 2001, but did not become part of the Hartland line after all. One of them depicts "Walter," a Quarter Horse gelding owned by Don Blazer, author of books on horse training and a horse advice column in rural newspapers. Walter, the real horse, attended the 2001 Jamboree.

After Hartland's plan fell through, the sculptor released the "Walter" mold herself late in 2002: 500 in Walter's black bay color, and 250 in each of four other colors, including palomino. The "Walter" mold is trotting, and was manufactured in China. The medium is resin, which had been Hartland's plan. At 7.75" tall and 12.5" long, he is scaled a little smaller than the Hartland 11" series.

The original price was $65. (Mass production holds down the prices on resin horses, making them more affordable than "artist resins," which are typically limited to editions of 50 or 100 because their cold-cast molds break down.) In February 2007, "Walter" was $35 and the four other colors, called "Change of Hart," were $50. In June 2011, "Walter" and the four "Change of Hart" colors were half-price ($25) for a limited time, and available in both glossy and matte finishes. The "Walter" mold was also available in custom colors, pre-painted by the sculptor, for $50.

The two other shapes of resin horses that almost became part of the Hartland line were a jogging stock horse mare and a trotting young Arabian mare. Read more in the "History" chapter.

## F & G

- "Seabiscuit" model

The 3.5" tall, resin model of Seabiscuit was sold in a box with "Hartland Collectibles" and "JF Sports Marketing" printed on it. It was produced in China. Consumers could not order the model from Hartland. Instead, it was auctioned on eBay by numerous distributors, starting in July 2003. Winning bidders often paid $15 or more in the first weeks, but less as time went on. It was priced at $1.99 on eBay in December 2009. Suggested values are: NM horse: $4; NM horse with NM box: $5; New in unopened box: $6.

A side panel of the box states that Hartland has manufactured horse replicas for over 50 years, but this model is no descendent of them. The real Seabiscuit was bay, but the model is reddish chestnut. The model's gait is closer to the fast trot of a harness racer than the gallop of a Thoroughbred race horse. "Seabiscuit" was evidently the result of a licensing arrangement in which the Hartland name was used.

The back of the box includes information such as Seabiscuit's date of birth (May 23, 1933), major owner (Charles S. Howard), and career statistics and honors. Seabiscuit won 33 races in 89 starts, and defeated War Admiral in "The Match of the Century" in 1938.

The namesake model was released shortly before the movie *Seabiscuit* opened in theatres on July 25, 2003. The movie was based on the excellent book, *Seabiscuit: An American Legend* by Laura Hillenbrand (Random House, 2001). Foaled in 1933, Seabiscuit was "Horse of the Year" in 1938, and captured the imagination of the public during the Great Depression.

**Honorable Mention.** This chapter is dedicated to a horse I met and one I just admired. Duster was a buckskin gelding who, I was told, had jumped at Madison Square Garden. When I met him in 1990, he was retired to a hobby farm near De Kalb, Illinois. Grazing apart from the other horses, he was socially self-sufficient and sort of above herd politics. When it came to people, he understood what was expected, and behaved to please. Among horses, Duster was what I call "a good citizen," a horse of the highest order!

Ledgeview's Asset was a black Saddlebred. I never met him, but admired him from the grandstand at horse shows in Wisconsin in the late 1970s-early 1980s. He was already an older horse then, but still won blue ribbons with his young riders. I never saw a horse trot with such an even cadence. Ledgeview's Asset kept perfect time!

## THE SPECIAL RUNS

**Holiday Models.** There were six holiday models in five years because 2001 had two – or should we say four? In 2001, "Noel" was sold on eBay, and "Sugarplum" was sold normally. Both were 9" models, and both purchases included a bonus item: a Tinymite. A pale gold Tinymite, "Glisten," came with "Noel." A black Tinymite, "Lump of Coal," came with "Sugarplum." The two Tinymites are not shown in the table. Most of the holiday items were numbered editions, but "Glisten," "Lump of Coal," and "Sugarplum" were not numbered.

In 2003 and 2005, the holiday item was a set of five Tinymites; they were numbered on a card that came in their package. In 2002 and 2004, the holiday model was 9" size, and edition-numbered on the belly, like "Noel" from 2001.

**HOLIDAY MODELS**
**NUMBERED EDITIONS ( MOSTLY )**

| Holiday | Model | Qty. |
|---|---|---|
| 2001 | "Sugarplum"<br>9" Polo Pony, cinnamon dapple<br>883-06 | 150 |
| 2001 eBay | "Noel"<br>9" Tennessee Walker, dapple grey w/white socks<br>887-01 | 85 |
| 2002 | "Snow Angel"<br>9" Five-Gaiter, dapple grey<br>881-06 | 62 |
| 2003 | "Frankincense"<br>set of five copper Tinymites<br>TM-5C | 30 sets |
| 2004 | "Heart of Gold"<br>9" Polo Pony, metallic gold bay<br>883-16 | 85 |
| 2005 | "Christmas Chocolates"<br>set of five brown Tinymites<br>TM-55 | 30 sets |

**Holiday Models** – The holiday models from 2001-2005 were 9" series models or Tinymites. Most were numbered editions. An example of the edition notation is: 052.

**CLUB MODELS**
**NUMBERED EDITIONS ( EXCEPT ONE )**

| Year | Model | Qty. |
|---|---|---|
| 2001 | "Chart the Course"<br>9" Polo Pony, bay appaloosa<br>883-01 | 270 |
| 2002 | "Wave the Banner"<br>9" Five-Gaiter, chestnut pinto<br>881-04 | 130 |
| 2003 | "Steadfast Hart"<br>11" Arabian, dapple grey<br>901-11 | 72 |
| 2004<br>Breed series option | "Cinn"<br>9" Polo Pony, metallic copper with white mane and tail<br>883-17 | 85 |
| 2004<br>Rider series option | "Buckeye"<br>9" Semi-Rearing, pale buckskin<br>850-B | 75 |
| 2005 | "Cuppa Joe"<br>9" Semi-Rearing, coffee dun<br>850-CD | 100 |
| 2006<br>Breed series option | "Lion Hart"<br>9" Tenn. Walker, bay with three socks<br>887-07 | 50 |
| 2006<br>Rider series option | Chief Thunderbird Tree<br>Unpainted Indian and accessory parts; no item (catalog) number, and was not edition-numbered | many |

**Club Models** – The club models were numbered editions. An example of the edition notation, hand-written in ink on the model's underside is: 062.

**eBay Specials.** Besides the 2001 eBay Holiday Edition, "Noel," one other production model was sold only on eBay: the prancing, 11" series Buckskin Appaloosa (902-03). An edition of 100, he debuted on eBay in late August 2001. He (or she) was not edition-numbered or named.

**Club Models.** The Hartland Horses Club (or Hartland Western Club – the two names were used almost interchangeably) ran for six years, but there were eight club models because in 2004 and 2006, collectors could choose between two models – or pay more and get both. The choice was: a breed-series horse or a rider-series item. The club models were numbered editions, except the 2006 rider-series item was not numbered.

Each membership also included a small item. For the first four years, it was a lapel pin. In 2005, it was a rifle with Hartland logo. In 2006, it was a Dale Evans hat key chain with Hartland logo.

**Jamboree Special Models.** For four years, 2001-2004, the Jamboree offered a ticketed (pre-ordered) special run from Hartland Collectibles, L.L.C. The Jamboree ordered the models from Hartland and sold them. In 2001 and 2002, extra models were produced, and could be bought after the event (from Horse-Power Graphics, Inc., the Jamboree host). In 2003 and 2004, though, the editions were limited to 50, and were sold out in advance.

**JAMBOREE SPECIAL MODELS**
**TICKETED ( PRE-ORDERED )**
**THEY HAD NAMES, BUT WERE NOT NUMBERED EDITIONS.**

| Dates | Model | Qty. |
|---|---|---|
| 2001 | "Tom-Tom"<br>11" Quarter Horse mold in black pinto, 902-01 | 250 |
| 2002 | "Jammer Time"<br>9" Polo Pony in buckskin overo pinto, 883-08 | 100 |
| 2003 | "TinyMite Draft Horse Set"<br>set of five Tinymite Draft horses, each painted a different and realistic color, TM-5D | 50 sets |
| 2004 | "The Tiny Five"<br>set of five Tinymites; five breeds, each in a different and realistic color, TM-51 | 50 sets |

**Jamboree Specials** – The "Jamboree special models" were not a mystery: They could be pre-ordered. They had names, but were not numbered editions.

**Surprise Jamboree Specials.** The Jamboree also featured "surprise specials," models that Hartland unveiled and sold at the Jamboree. The table below lists only the surprise specials that sold out at the Jamboree. Some models that debuted at Jamboree were later available at the company website, so count as regular runs. None of the Jamboree specials – either ticketed or unannounced – were edition-numbered.

| SURPRISE JAMBOREE SPECIALS THAT SOLD OUT AT THE EVENT | | |
|---|---|---|
| Event | Model | Qty. |
| Jamboree 2003 | 9" Mustang, blue roan, 886-06; 22 sold out at Jamboree, and 3 were HPC gifts, Summer 2003<br>886-06 | 25 |
| Jamboree 2003 | Blue Roan Tinymite Draft Horse<br>TM-1D | 35 |
| Jamboree 2004 | 9" Tennessee Walker, dapple grey with black points<br>887-05 | 25 |
| Jamboree 2004 | "The Might-y Five"<br>set of five Tinymites: five breeds, each a different color<br>TM-52 | 50 sets |

**Surprise Jamboree Specials** – Counted as "surprise Jamboree specials" are models that were unveiled at, and sold out during, the Jamboree. They were not numbered editions.

**Show Special Models.** The show special models were sold by member shows of the Hartland Horses Show Producers Club. The Jamboree was one of the member shows. The show special models were not edition-numbered.

| SHOW SPECIAL MODELS NAMED, BUT NOT NUMBERED | | |
|---|---|---|
| Year | Model | Qty. |
| 2001 | "Silver Sultan"<br>11" Arabian, bay with dapples<br>901-01 | 305 |
| 2002<br>1st half | "Copper King"<br>11" Arabian, chestnut pinto<br>901-07 | 76 |
| 2002<br>2nd half | "Bronze Ruler"<br>11" Arabian, bay pinto<br>901-08 | 46 |
| 2003 | "Black Tie"<br>9" Polo Pony, black overo pinto<br>883-09 | 78 |

**Show Specials** – In 2001-2003, there were four show special models: three 11" Arabians and one 9" Polo Pony. They had names, but were not numbered editions.

**Models Sold by Mail-in Drawing.** Four models were sold only to club members whose names were drawn after they mailed their contact information to the company. The models were two Mustangs and two 11" series Arabians.

| MODELS SOLD BY MAIL-IN DRAWING (NOT NUMBERED EDITIONS) | | |
|---|---|---|
| Date | Model | Qty. |
| Aug. 2001 | 9" Mustang<br>Buckskin Pinto with Black Tail<br>886-03B | 21 |
| Jan. 2002 | 9" Mustang<br>Black (Pearled)<br>886-04 | 11 |
| Dec. 2002 | 11" Arabian<br>Black Pinto with Star Face<br>901-10W | 15 |
| Aug. 2005 | "Hart of Blues" 11" Arabian<br>Dark Blue w/ White Mane and Tail,<br>901-HB | 30 |

**Sold by Drawing** – Two Mustangs and two 11" Arabians were sold by mail-in drawing. They were not numbered editions, but one was named.

**Noble Horse Series and "Cocoa."** The two models in the Noble Horse Series were numbered and signed by the artist who painted them, Sheryl Leisure. An example of the numbering is: 17/20. "Cocoa" has the same style of edition notation, but is not signed.

| NOBLE HORSE SERIES AND A MISCELLANEOUS NUMBERED EDITION | | |
|---|---|---|
| Timing | Model | Qty. |
| 1st model in NHS -- June 2006 | "Viceroy"<br>11" Quarter Horse, dappled buckskin with black points<br>902-11 | 20 |
| 2nd model in NHS -- Jan. 2007 | "Volant"<br>9" Polo Pony, dappled palomino with green wraps<br>883-19 | 15 |
| Dec. 2007 | "Cocoa"<br>9" Semi-Rearing, chocolate bay<br>850-CB | 20 |

**Noble Horse Series** – In the Noble Horse Series (2006-2007), the two models were both numbered and signed by the artist. A third horse, "Cocoa," was numbered, but unsigned. An example of the notation is: 04 / 20 (4th piece in a run of 20).

## Eying the Rider Series Animals

All six of the 9" rider-series horses produced in 2001-2007 have tri-color eyes.

The Pale Buckskin Semi-Rearing horse (March 2004) has an eye that is predominately black and white. The front and back corners of the eye are white. There's a black circle with brown "corona"; in other words, a large black pupil with a small brown iris surrounding it.

The process likely began with adding brown shading to the eye and the area surrounding it. Then, the entire eye was sprayed white using a mask to define the painted area. Then, medium reddish brown (burnt orange) color was sprayed on, using a mask. Then, a black circle, also masked, was sprayed over most of the brown.

Brown predominates in the eyes of the Semi-Rearing Coffee Dun, March 2005. First, black or dark brown shading was sprayed over the eye region. Both the front and back eye corners are pale peach. A dark, taupe-brown iris looks hand-painted. Then, a round, black pupil was hand-painted without a mask.

The Coffee Dun Chubby horse from April 2005 has the same eye style except that the corners are white.

The Semi-Rearing horse in Chocolate Bay, December 2007, has an all-black eye except for a white front corner added with a brush.

Trigger from the Roy Rogers set had a small amount of brown shading sprayed over the greater eye area. A medium, taupe-brown iris occupies most of the eye. The white in the front and back corners looks free-hand brushed in, and the circular, black pupil also looks hand-brushed without a mask.

Buttermilk from the Dale Evans set has eyes like Trigger's except that the brown iris is a lighter color: light taupe-brown.

Bullet, the German Shepherd, has black shading over his face. Then, white was sprayed in his eyes; next, a medium, reddish brown iris; and then, a round black pupil. Trigger, Buttermilk, and Bullet were painted in China.

## Eyes on the Horizon: 11" Series

With his head held high, the 11" Arabian appears to have his eyes on the horizon. He and the 11" series Quarter Horses have the largest eyes in Hartland line since 2000, but they are still only about ⅜" high and ¼" wide from front to back corner. Hartland could have simply painted the eyes black or brown, but it chose to fit a lot of detail onto a small, curved surface, and kept tinkering with the design. The 21 models in the 11" series evolved through several different designs, described here in order of model release.

The first 11" series model was the 2001 show special, "Silver Sultan," the Dappled Bay Arabian. My example from May 2001 has a black eye with white-dot highlight in the black area.

"Tom-Tom" the Black Pinto (QH mold) from June 2001 has the black eye with white dot, but adds a white crescent near, but not at, the back corner of the eye. It is not a "white corner"; the corner of the eye is black! The crescent is a curved, vertical "stripe" from top to bottom of the eye. It makes the eye appear to be looking forward. This style of eye is also found on both of the Arabs from June 2001, Silver-Gray and Black, and on my darker version of "Silver Sultan" from a show in November 2001.

The July 2001 Copper Chestnut Arab has the same pattern of white (white dot and crescent), but no black color! Instead, dark brown shading was sprayed over the greater eye area (the eye and the area surrounding it), and the white dot and crescent were added to that.

The first type of tri-color eye debuted on the Buckskin Appaloosa (QH mold), in September 2001. On him, the eye itself is mostly black, but has a brown crescent near the back corner, white in the back corner, and a white dot (highlight) in the black area. The brown and white areas were hand-applied (with a manual brush, not an airbrush). As a preliminary step, black shading was sprayed over his greater eye area. That first procedure was used on most of the 11" series. Three models from October 2001 have that same eye type: Arabians in Palomino and Buckskin Grey, and the Chocolate Bay Quarter Horse, although his brown crescent is a lighter, brighter shade of brown: burnt orange.

The eye on the Blue Roan Appaloosa (QH mold), December 2001, is mostly black, but has two brown crescents bracketing the black area, and white painted in the front and back corners. So, from left to right, there are five sections of color in each eye, but there is no dot (highlight).

The Palomino Quarter Horse, February 2002, looks like he's wearing eye liner because of the black paint intentionally sprayed over the greater eye area. His eye style is mostly black with a thin, brown crescent toward the back, and a white back corner painted in by hand. There is no highlight (dot). The same eye is seen on "Copper King," the Chestnut Pinto Arabian.

Models from April through July 2002 have the white dot highlight along with the black eye with thin brown crescent near the back, and a white back corner. They are: the Flaxen Chestnut Arabian, April 2002; "Bronze Ruler" the mid-2002 Bay Pinto Arabian; and the Slate-Gray Overo Pinto (QH mold), June 2002. The Bright Bay Quarter Horse, July 2002, is the same, but his brown crescent may be burnt orange, rather than brown.

For the Black Pinto Arabian with Star Face, December 2002, the highlight was omitted. His eye type is: black to the front, with a brown crescent near the back, and a white back corner.

In January 2003, "Steadfast Hart," the Dapple Grey Arab, has an entirely different style of tri-color eye. His has a peach crescent at the front of the eye (a whole crescent, not just a peach corner), a burnt-orange crescent next, and the rest of the eye is black with no highlight. He appears to be looking sideways!

The Black Appaloosa (QH mold), released in January 2003, has yet another eye type. His eyes are mainly black and peach. It appears that the peach color was hand-painted in, and then a brown (burnt orange) iris (unmasked circle) was added, and then a black pupil within the brown iris. The brown surrounds the black pupil on three sides: all except the top. There is no highlight.

There was a two-and-one-half year interval before the next 11" series model, "Hart of Blues," the Blue Arabian. He had black sprayed to the general eye area of the face, and then a large, white crescent painted in the front corner. There is no dot highlight.

Lastly, "Viceroy," the Dappled Buckskin QH, June 2006, has the most complex eye of them all. Slightly darker shading was sprayed over his general eye area. The front corner of the eye is peach (or pink), and the back corner is white. He has what I call a "donut" eye. It has a black ring with a brown ring painted inside it, and a rectangular black center. The colors were all hand painted, and the eye looks glossed.

## The Eyes Have It: 9" Breed Series

Without a "neigh" of dissent, the "eyes" have it: The models since 2000 indisputably have more detailed eyes than the models from any previous Hartland era. In the 9" horse breeds series, there were 37 production models, and the eyes evolved through several designs from 2001-2007. On all but one of those 9" models, the eye styles have something in common that separates them from most of the 11" models from 2001-2007. The 9" models do not have darker shading on the head around the eye area to set off the eye itself. Otherwise, their eyes are as complicated as on the 11" series models.

(The descriptions that follow are based on the single example I own of 35 of the models, plus accounts from other collectors for the other two models, Mustangs in pearl black and in buckskin pinto with black tail. The eyes of a model you own could differ slightly from these descriptions.)

The 9" series horses started out with a black eye with white-dot highlight and a white back corner of the eye. (A dot highlight, on any Hartland model, is always in the black-colored part of the eye.) The black color was airbrushed on whereas the white areas were typically added by hand. This scheme is found on most of the models released from February 2001 – October 2001, and on three models after that. The first releases, the Bay Appaloosa Polo Pony (February) and Palomino Five-Gaiter (March) have it. It is also seen on summer and fall 2001 models: Polo Ponies in Flecked Grey (June), Black Appaloosa (July), and Liver-Chestnut Pinto (November); on Five-Gaiters in Black with white points (October) and in Rose Grey (Dapple Grey with Bay Shadings, February 2002), and on two Mustangs: Buckskin Pinto Mustangs with black tail (August 2001) and Pearl Black (January 2002).

An exception is that two Mustangs released in June 2001 had an eye that was black with a white corner in front, not in back; and their eyes did not have a dot highlight. They are the Chocolate Bay and Black-Gray Pinto Mustangs. The white corner, combined with their rearing pose, gives them a wild-eyed look.

In November 2001, the first tri-color eye appeared on a 9" series horse, the Cinnamon Dapple Polo Pony. On this model, the eye is mostly black, but has a vertical crescent of brown (burnt orange) toward the back of the eye, and white in the back corner. The brown area was painted on as a circle, and then it was mostly covered with black, leaving a brown crescent. There is no dot highlight. Three more Polo Ponies also had the black eye with brown crescent behind it, and then a white back corner and no dot highlight in the black area: Dapple Grey (March 2002), Palomino (September 2002), and Black Overo Pinto (June 2003). Another model with a similar eye is the Five-Gaiter in Dapple Grey (November 2002), whose brown crescent is copper brown.

Most of the models released from mid-November 2001 through August 2002 had a tri-color eye with a dot highlight in the black area. Otherwise, they look the same as the eyes in the group just described; that is, they have a brown crescent behind the black area, and a white back corner. The first model like this (with dot highlight) is the Dapple Grey Tennessee Walker, "Noel" (November 2001). He was followed by the Black Tennessee Walker and Palomino Pinto Mustang in December 2001; Five-Gaiter in Chestnut Pinto (January 2002); Copper Chestnut Tennessee Walker and Caramel Bay Five-Gaiter in July 2002, and Buckskin Pinto Polo Pony in August 2002. The Blue Roan Tennessee Walker, June 2002, has one eye like this while his left eye is "blue" in keeping with his apron-face white marking. (It is a donut with a black outer ring, light blue middle ring, and a black center.)

The Dapple Grey Tennessee Walker with black points released in October 2004 also has this same eye, lending credibility to the idea that he was created by adding black points to leftover "Noel" models. That's because new eye types began a year before him.

October 2003 marked the beginning of new eye variations in the 9" series. The Blue Roan Mustang (October 2003) has a peach front corner, and the rest of the eye is black, with no dot highlight. The eye was glossed.

In 2004, the Metallic Copper Polo Pony (March 2004) had a white front corner that is a big, white crescent. The rest of the eye is black with no dot highlight. This same eye style is found on the Metallic Gold Polo Pony, November 2004.

In December 2004, the five "sample run" models, so-called because they were by a different painting service, all had very complicated eyes. The Bay Pinto Five-Gaiter has a white crescent in the front corner, and the rest of the eye is black except for a metallic gold "C" facing toward the back of the eye. It is a definite "C," not a donut.

The Buckskin Tennessee Walker has a pink front corner crescent, and the rest is black except for a metallic gold eyelash shape lying low in the black area with its ends turned up.

The Dun / Grulla Polo Pony has four colors in its eye. It has white in both the front and back corners. (The front corner is large enough to be called a white crescent.) There's a black circle, and on three sides of it (all but the top), there's a metallic gold crescent, and then a taupe brown crescent below that. It appears that the taupe-brown was painted first, then the gold, and then the black.

The Bay Roan Polo Pony has pink in the front and back corners, and the rest of the eye is black except for a thin line of metallic gold "outlining" the black area, but slightly inside it, so there is still a black border.

The Bay Pinto Polo Pony has white in both corners and a donut eye with a rounded center. The donut is formed by a large, black circular area, a metallic gold circle drawn inside it, and a black center.

That brings us to 2006. The Bay Tennessee Walker has a pink front corner and white back corner, and a "donut" eye that is black and golden brown (either copper-colored or burnt orange), and looks glossed. It has four colors in the eye! The Liver Chestnut Five-Gaiter and Red-Chestnut Pinto Polo Pony have the same eye, but have white in the front and back corners.

Even more detailed work went into the Palomino Polo Pony with green wraps, January 2007. It has the donut eye with pink in both corners, but it also is the only 9" breed-series model with dark shading around the eye. The shading was a preliminary step. It makes the eyes stand out more.

## Production Summary

| Hartland Horses' 2001-2007 Production Models (in Plastic) Grouped by Month and Year of Debut | | | | | |
|---|---|---|---|---|---|
| **Color** | **Shape** | **Item #** | **Date** | **Qty.** | **Orig. Price** |
| **2001** | | | | | |
| "Chart the Course" Copper-Bay Appaloosa | 9" Polo Pony | 883-01 | Feb. 2001 | 270 | $50* |
| Palomino (pearled)<br>Champagne (pearled) | 9" Five-Gaiter<br>7" TW Family | 881-01<br>684-01 | March 2001 | 150<br>150 | $36<br>$45 |
| Black Pinto | 7" TW Stallion | 684S-01 | April 2001 | 125 | $25 |
| "Silver Sultan" Bay w/ dapples | 11" Arabian | 901-01 | May 2001 | 305 | $45 |
| "Tom-Tom" Black Pinto<br>Flecked Grey, turquoise wraps<br>Chocolate Bay<br>Black-Gray Pinto<br>Silver-Gray (pearled)<br>Black (pearled) w/ white socks | 11" QH mold<br>9" Polo Pony<br>9" Mustang<br>9" Mustang<br>11" Arabian<br>11" Arabian | 902-01<br>883-02<br>886-01<br>886-02<br>901-02<br>901-03 | June 2001 | 250<br>50<br>100<br>50<br>50<br>100 | $50<br>$32<br>$32<br>$32<br>$36<br>$36 |
| Copper Chestnut<br>Black Appaloosa, golden wraps<br>Black (dappled & pearled) | 11" Arabian<br>9" Polo Pony<br>7" TW Family | 901-04<br>883-03<br>684-02 | July 2001 | 50<br>75<br>50 | $36<br>$34<br>$45 |
| Buckskin Pinto w/ Black Tail | 9" Mustang | 886-03B | Aug. 2001 | 21 | $32 |
| Buckskin Appaloosa | 11" QH mold | 902-03 | Sept. 2001 | 100 | $45 |
| Chocolate Bay<br>Black w/ white points (pearled)<br>Palomino (dappled)<br>Buckskin Grey | 11" QH mold<br>9" Five-Gaiter<br>11" Arabian<br>11" Arabian | 902-02<br>881-02<br>901-05<br>901-06 | Oct. 2001 | 50<br>50<br>56 (56)<br>50 | $36<br>$36<br>$36<br>$36 |
| "Sugarplum" Cinnamon Dapple<br>"Lump of Coal" Black<br>Liver-Chestnut Overo Pinto<br>"Noel" Dapple Grey (pearled)<br>"Glisten" Pale Gold | 9" Polo Pony<br>Tinymite (one)<br>9" Polo Pony<br>9" Tenn. Walker<br>Tinymite (one) | 883-06<br>TM-6B<br>883-04<br>887-01<br>TM-6G | Nov. 2001 | 150<br>50 of ea.<br>50<br>85<br>20 of ea. | $36<br>$3 ea.<br>$34<br>$45<br>N/Ap |
| Palomino Pinto<br>Black with white socks<br>Blue Roan Appaloosa | 9" Mustang<br>9" Tenn. Walker<br>11" QH mold | 886-05<br>887-02<br>902-07 | Dec. 2001 | 50<br>50<br>100 | $34<br>$36<br>$38 |

**Production Models in Chronological Order** – This table, and the ones on the following pages, list the production models for each year, with quantity produced and original price. The models are in chronological order, by year and month of release. If you bought each model when it first came out, the amounts in the far right column are the prices you paid. An asterisk (*) indicates that the price included shipping. Otherwise, shipping was extra. Items with "N/Ap" ("not applicable") were sold with another item. Two asterisks (**) note that in 2004, both club models purchased together were $80 including shipping.

| Hartland Horses' 2001-2007 Production Models (Continued: 2002-2003) | | | | | |
|---|---|---|---|---|---|
| **Color** | **Shape** | **Item #** | **Date** | **Qty.** | **Orig. Price** |
| **2002** | | | | | |
| "Wave the Banner" Chestnut Pinto<br>Black (pearled)<br>Buckskin | 9" Five-Gaiter<br>9" Mustang<br>7" TW Family | 881-04<br>886-04<br>684-03 | Jan. 2002 | 130<br>11<br>50 | $50*<br>$34<br>$45 |
| "Copper King" Chestnut Pinto<br>Bay Pinto<br>Black Pinto<br>Rose Grey (Dapple Grey w/ Bay Shadings)<br>Palomino (Dappled)<br>Dapple Grey w/ white wraps | 11" Arabian<br>7" TW Foal<br>7" TW Foal<br>9" Five-Gaiter<br>11" QH<br>9" Polo Pony | 901-97<br>684F-02<br>684F-03<br>881-03<br>902-04<br>883-07 | Feb. 2002 | 76<br>50<br>50<br>50<br>50<br>50 | $45<br>$12<br>$12<br>$34<br>$38<br>$34 |
| Flaxen, Liver Chestnut | 11" Arabian | 901-09 | April 2002 | 50 | $36 |
| Black Saddle w/ rifle hole<br>Brown Saddle w/ rifle hole<br>Blue Roan<br>Slate-Gray Overo Pinto / Paint | 9" accessory<br>9" accessory<br>9" Tenn. Walker<br>11" QH mold | --<br>--<br>887-03<br>902-08 | June 2002 | many<br>many<br>50<br>50 | $25<br>$25<br>$36<br>$38 |
| "Bronze Ruler" Bay Pinto | 11" Arabian | 901-08 | Mid-2002 | 46 | $38 |
| Bright Bay<br>Caramel Bay (dappled)<br>Copper Chestnut w/ white socks | 11" QH<br>9" Five-Gaiter<br>9" Tenn. Walker | 902-09<br>881-05<br>887-04 | July 2002 | 50<br>50<br>50 | $36<br>$34<br>$36 |
| "Jammer Time" Buckskin Pinto | 9" Polo Pony | 883-08 | Aug. 2002 | 100 | $36 |
| Dapple Grey<br>Pale Palomino with white wraps | 7" TW Stallion<br>9" Polo Pony | 684S-03<br>883-05 | Sept. 2002 | 50<br>35 | $25<br>$34 |
| "Snow Angel" Dapple Grey | 9" Five-Gaiter | 881-06 | Nov. 2002 | 62 | $30 |
| Black Pinto w/ Star Face | 11" Arabian | 901-10W | Dec. 2002 | 15 | $36 |
| **2003** | | | | | |
| "Steadfast Hart" Dapple Grey<br>Black Appaloosa w/ Dusty Hips | 11" Arabian<br>11" QH mold | 901-11<br>902-10 | Jan. 2003 | 72<br>50 | $50*<br>$34.50 |
| Brown Indian Saddle<br>Indian Accessories Set #1: Black, w/ spear | 9" accessory<br>9" accessories | --<br>-- | May 2003 | many<br>many | $15<br>$30 |
| "Black Tie" Black Overo Pinto | 9" Polo Pony | 883-09 | June 2003 | 78 | $35 |
| Bay (with Pearled undercoat) | 7" TW Family | 684-04 | Aug. 2003 | 25 | $38 |
| "TinyMite Draft Horse Set"<br>Blue Roan Draft Horse<br>Blue Roan | Tinymites<br>Tinymite (one)<br>9" Mustang | TM-5D<br>TM-1D<br>886-06 | Oct. 2003 | 50 sets<br>35<br>25 | $32.50<br>$20<br>$30 |
| "Frankincense" Copper color | Tinymites | TM-5C | Dec. 2003 | 30 sets | $32.50 |

| HARTLAND HORSES' 2001-2007 PRODUCTION MODELS<br>CONTINUED: 2004-2007 | | | | | |
|---|---|---|---|---|---|
| **Color** | **Shape** | **Item #** | **Date** | **Qty.** | **Orig. Price** |
| **2004** | | | | | |
| "Cinn" Metallic Copper<br>"Buckeye" Pale Buckskin | 9" Polo Pony<br>9" Semi-Rearing | 883-17<br>850-B | March 2004 | 85<br>75 | $50**<br>$50** |
| Hat Key Chain with tag, no logo | 9" accessory | -- | June 2004 | 100 | N/Ap |
| "The Tiny Five"<br>"The Might-y Five"<br>Dapple Grey with black points | Tinymites<br>Tinymites<br>9" Tenn. Walker | TM-51<br>TM-52<br>887-05 | Oct. 2004 | 50 sets<br>50 sets<br>25 | $32.50<br>$32.50<br>$30 |
| "Heart of Gold" Metallic Gold Bay | 9" Polo Pony | 883-16 | Nov. 2004 | 85 | $28 |
| Bay Pinto<br>Buckskin<br>Dun / Grulla<br>Bay Roan<br>Bay Tobiano Pinto | 9" Five-Gaiter<br>9" Tenn. Walker<br>9" Polo Pony<br>9" Polo Pony<br>9" Polo Pony | 881-07<br>887-06<br>883-10<br>883-15<br>883-13 | Dec. 2004 | 104<br>58<br>39<br>44<br>44 | $19.99<br>$19.99<br>$19.99<br>$19.99<br>$19.99 |
| **2005** | | | | | |
| "Cuppa Joe" Coffee Dun<br>Rifle w/ Hartland logo | 9" Semi-Rearing<br>9" accessory | 850-CD<br>-- | March 2005 | 100<br>100 | $45*<br>N/Ap |
| "Java Joe" Coffee Dun<br>Roy Rogers Saddle, square fender | 9" Chubby S/W<br>9" accessory | 820-CD<br>-- | April 2005 | 56<br>many | $35<br>$20 |
| "Hart of Blues" Dark Blue | 11" Arabian | 901-HB | Aug. 2005 | 30 | $40 |
| "Christmas Chocolates"<br>Roy Rogers and Trigger<br>Dale Evans, Buttermilk, Bullet | Tinymites<br>9" rider set<br>9" rider set | TM-55<br>#806<br>#802 | Dec. 2005 | 30 sets<br>2,500<br>2,500 | $32.50<br>$50<br>$50 |
| **2006** | | | | | |
| "Lion Hart" Bay<br>Chief Thunderbird Tree (white)<br>Hat Key Chain w/ Hartland logo | 9" Tenn. Walker<br>9" rider series<br>9" rider series | 887-07<br>--<br>-- | Jan. 2006 | 50<br>many<br>100-150 | $55<br>$55<br>N/Ap |
| Flaxen, Liver Chestnut<br>Red-Chestnut Overo Pinto<br>Bay Roan<br>Chestnut Roan | 9" Five-Gaiter<br>9" Polo Pony<br>7" TW Family<br>7" TW Family | 881-09<br>883-18<br>684-BR<br>684-CR | May 2006 | 25<br>25<br>25<br>25 | $30<br>$30<br>$38<br>$38 |
| "Viceroy" Buckskin (dappled) | 11" QH | 902-11 | June 2006 | 20 | $75 |
| **2007** | | | | | |
| "Volant" Palomino (dappled) w/ green wraps | 9" Polo Pony | 883-19 | Jan. 2007 | 15 | $50 |
| Indian accessories set #2:<br>Black, w/ saddle (6 pc)<br>Indian accessories set #3:<br>Brown (6 pc)<br>Rifle: brown & silver | 9" accessories<br><br>9" accessories<br><br>9" accessory | --<br><br>--<br><br>-- | Aug. 2007 | many<br><br>50 sets<br><br>50 | $30<br><br>$30<br><br>N/Ap |
| "Cocoa" Chocolate Bay | 9" Semi-Rearing | 850-CB | Dec. 2007 | 20 | $45 |

| PRODUCTION EDITIONS BY YEAR: HORSE ITEMS & ACCESSORY ITEMS, 2001-2007 (QUANTITY OF CATALOG ITEMS & TOTAL PRICE) | | | |
|---|---|---|---|
| **Year** | **Horse Items: Quantity & Total Price** | **Accessories: Quantity & Total Price** | **Total Price for Year(s)** |
| 2001 | 28 items, $993 | None | $993 |
| 2002 | 21 items, $719 | 2 items, $50 | $769 |
| 2003 | 8 items, $272.50 | 2 items, $45 | $317.50 |
| 2004 | 11 items,$322.95 | 1 item, N/Ap | $322.95 |
| 2005 | 6 items, $252.50 | 2 items, $20 | $272.50 |
| 2006 | 6 items, $266. | 2 items, $55 | $321 |
| 2007 | 2 items, $95. | 3 items, $60 | $155 |
| **Total** | **82 horse items, $2,860.95** | **12 items, $230** | **7 Years: $3,090.95** |

**Collection Value** – If you bought one of each of the 82 new horse catalog items when they first came out in 2001-2007, you paid about $2,860.95. The average price was about $35. The 12 new accessory items totaled $230, and averaged $19. The grand total price was $3,090.95, which does not count shipping (except for "club" models).

**A Hectic Pace.** Models were released at a hectic pace in 2001 and 2002, and then, at a slower rate in 2003-2007. The number of different horse items released each year, and their value (original price), is summarized in the table above.

**Shipping.** The shipping was almost always in addition to the price. It was: $6.95 for one model (whether 7", 9", or 11" series) or for a three-piece, 7" series family; $10.95 for two models; and $14.95 for three. There were some exceptions. For club membership items, shipping was included in the price. If you bought all five of the "sample run" models in December 2004, you paid only $11.95 shipping. The shipping for "Cocoa" in December 2007, the final Hartland horse to date, was $9.95.

**The Most Numerous Models.** The Roy Rogers and Dale Evans horse-and-rider sets, editions of 2,500 each, were the only "mass produced" new releases in plastic. All other plastic horses numbered fewer than 350 pieces. Most editions were limited, so how many more could have been sold if given the chance, is not known.

The best-selling horses under 350 pieces were:

1. "Silver Sultan," the 11" Arabian that was the 2001 show special model – 305 made
2. "Chart the Course," the Polo Pony that was the 2001 club model – 270
3. "Tom-Tom," the 11" Paint (black pinto Quarter Horse) that was the ticketed, special run at the 2001 Jamboree – 250

Six other horse items turned out to be editions over 100. They are:

1. The Palomino (Pearled) Five-Gaiter – 150
2. The 7" Champagne Tennessee Walker Family – 150
3. "Sugarplum," the Polo Pony that was a 2001 holiday model – 150
4. "Wave the Banner," the Five-Gaiter that was the 2002 club model – 130
5. The Black Pinto 7" Tennessee Walker Stallion – 125
6. The Bay Pinto Five-Gaiter sample run – 104

Five horses were limited to editions of 100:

1. The Chocolate Bay Mustang
2. The 11" Arabian in Black (Pearled)
3. The Blue Roan 11" Appaloosa (QH mold)
4. "Jammer Time," the 2002 Jamboree special model: a buckskin overo-pinto Polo Pony
5. "Cuppa Joe," the Semi-Rearing horse with saddle that was the 2005 club model

**Variety.** The production models were a large and varied group, and very well painted!

# PART 2
# GIFT RUN MODELS

Gift-run models were very small editions of models that Hartland only gave as gifts and did not sell. Part 2 includes three types of gift runs.

1. Gifts to Jamboree Helpers. In 2001 and 2002, the judges and other volunteers at Jamboree were gifted with a Hartland model different from any models that were ever for sale.
2. Prizes for Hoof Pick Contests. Most gift horses were awarded as prizes for the Hoof Pick Contest (abbreviated "HPC"). Nearly every issue of the company's quarterly newsletter had a Hoof Pick Contest. The contest was always the same. Readers were to locate the tiny image of a hoof pick – about one-quarter-inch long – embedded somewhere in the newsletter, and report, by mail, the page on which it was found. For each contest, between three and six winners were randomly drawn from the entries with the correct answers. Almost every contest had a different prize model run, and readers entered the contest without knowing what the prize model run would be. Those models were not ever for sale, with one exception. There was one production run from which three pieces were set aside for an HPC prize. (It was the Blue Roan Mustangs. See Chapter 2.)
3. Gifts to Member Shows. "Member Shows" were model horse shows that joined the Hartland Horses Show Producers Club. Member shows that met certain requirements were given a few models to use as door prizes or show trophies, and the shows were allowed to raffle one model for the show's benefit.

Some of the gifts to member shows were actually Hoof Pick horses. However, the models Hartland donated as show awards included unique models, and there was at least one small run that was used solely as show awards. The unique models are in Part 3, but the small run is included here.

**Values for Gift Run Models.** The gift runs are quite attractive, and most of them were editions of only three to six pieces. Their original recipients are aware of their scarcity, and likely to maintain them in near mint (if not perfect) condition. If or when the time comes to sell them, they are likely to be circulated back into the collecting community. They won't end up in a garage sale or a toy box.

Since the company gave the gift run models away, there is no initial price to factor into their value, but models from such small editions are usually worth about the same as unique (test) models in their same scale. The gift runs seldom come up for sale, but in December 2010, a 9" Thoroughbred in silver-gray sold on eBay for $127.50. Values are suggested for the four gift runs that were larger than 10 pieces.

**Spin-offs.** It appears that four gift runs were spin-offs of production models. Two of the gift runs are metallic gold bay Polo Ponies that resemble "Heart of Gold," the production run, but their wraps are a different color. They may have been made by simply adding red or blue paint to the white wraps on "Heart of Gold."

The Polo Pony gift run in Cinnamon Dapple Bay resembles the "Sugarplum" production models, and may have been created by adding black points to leftover "Sugarplums." Likewise, the gift-run 9" Five-Gaiter in dapple grey like "Snow Angel," but with black mane and tail, was probably the result of adding black paint to the surplus production of "Snow Angel." Hartland was clever about building the capacity for color alterations into the model colors.

The 9" Glowing Bay Thoroughbred *(upper)* is alert, but looking in the wrong direction to find the three hoof picks in the composite image of horse equipment. He's from a gift edition, 873-BG. The equipment photo, from 1988, was an unintentional multiple exposure. *Model photo, courtesy of Elaine Boardway.*

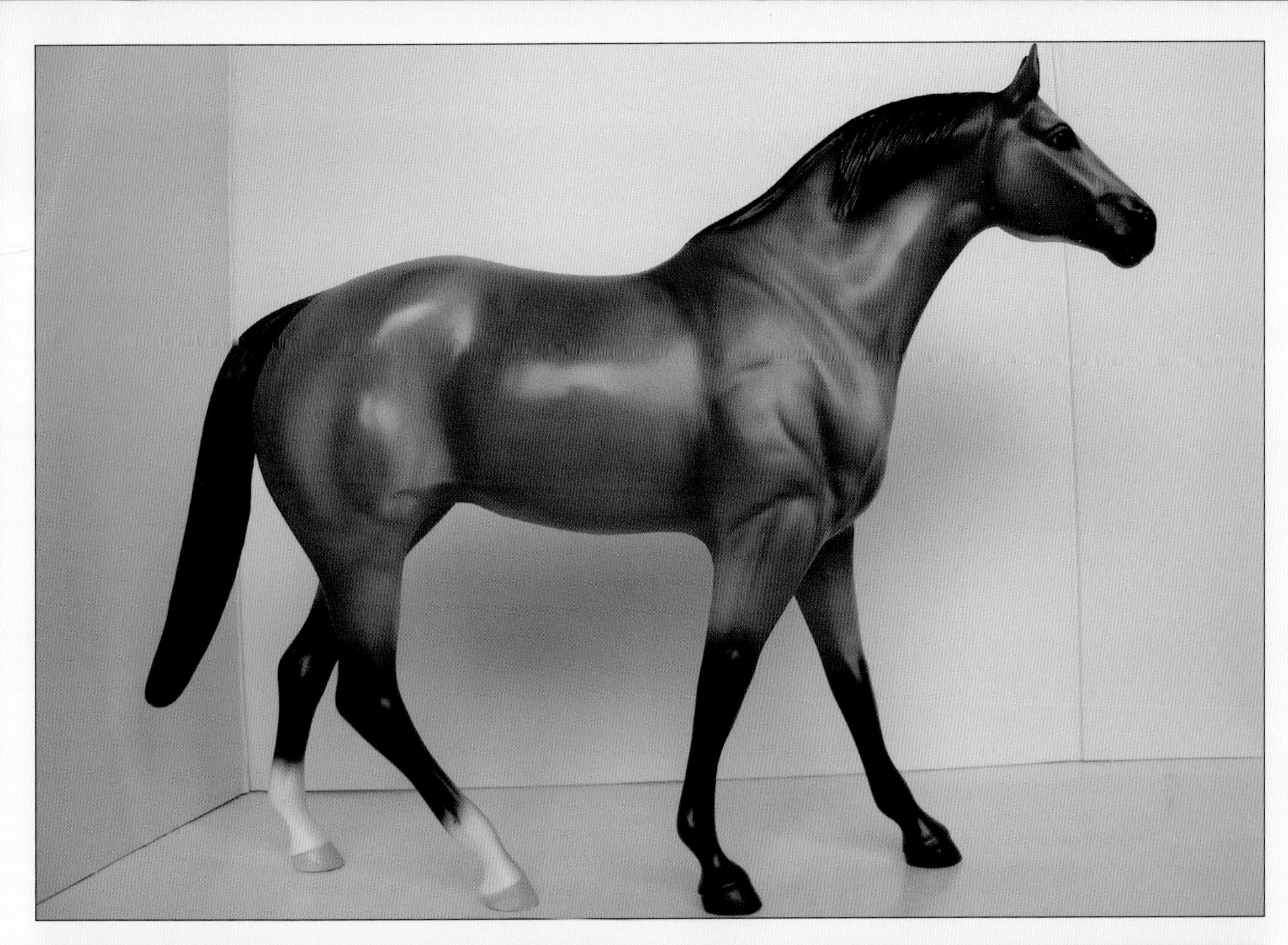

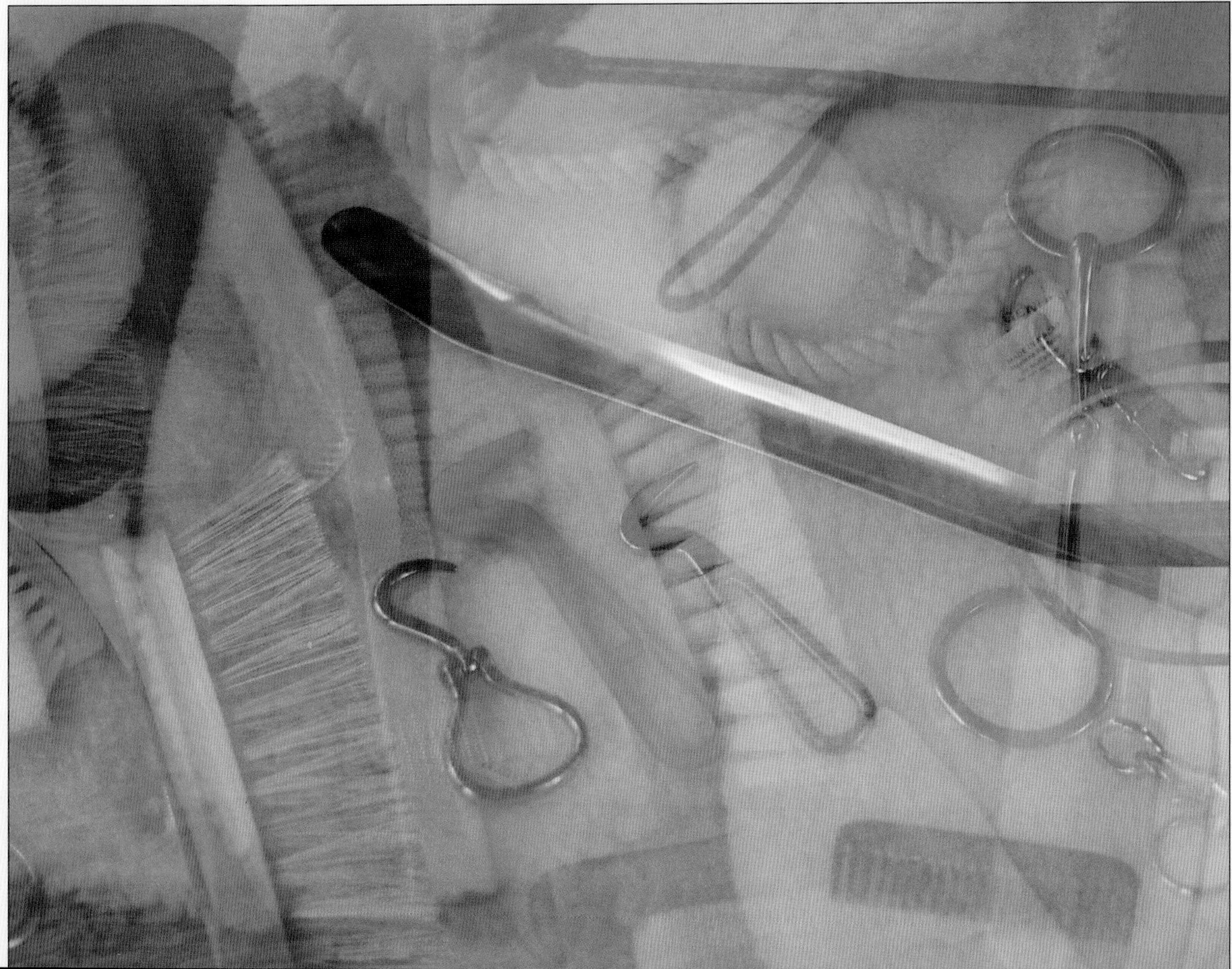

| GIFT RUN HORSES | | | | |
|---|---|---|---|---|
| **Breed & Mold** | **Color (and Name, if Applicable)** | **Item #** | **Qty.** | **Occasion** |
| 9" Thoroughbred<br>9" Thoroughbred | Silver-Gray with black points<br>Bay, glowing | 873-SG<br>873-BG | 5<br>3 | HPC, Summer 2002<br>HPC, Spring 2003 |
| 9" Five-Gaiter<br>9" Five-Gaiter | Dapple Grey with black mane/tail Black (Dappled, Pearled) | 881-08<br>881-BP | 5<br>3 | HPC, Fall 2002<br>HPC, Summer 2005 |
| 9" Mustang<br>9" Mustang | Buckskin Pinto with White Tail<br>Bay Pinto (Pearled) | 886-03W<br>886-BP | 29<br>3 | Gift to Jamboree helpers, June 2001<br>Show Trophy Run, 2002 |
| 9" Polo Pony<br>9" Polo Pony<br>9" Polo Pony<br>9" Polo Pony | Bay Overo Pinto<br>Metallic Gold Bay with red wraps Metallic Gold Bay w/ blue wraps<br>Cinnamon Dapple Bay | 883-BP<br>883-GR<br>883-GB<br>883-CDB | 6<br>3<br>3<br>3 | HPC, Winter 2003 – 3; Member Show Gifts – 3<br>HPC, Summer 2004<br>HPC, Fall 2004<br>HPC, Winter 2005 |
| 11" Arabian | Black Pinto with no star on face | 901-10B | 35 | Gift to Jamboree helpers, August 2002 |
| 7" TWH Stallion<br>7" TWH Stallion<br>7" TWH Stallion | "Ascot," Pearled Sorrel ("Champagne") with light mane/tail<br>"Dapper," Copper Chestnut Pinto<br>Bay Tobiano Pinto | 684S-02<br>684S-05<br>684S-04 | 18<br>12<br>5 | HPC, Spring 2001 – 5; Member Show Gifts – 13<br>HPC, Winter 2002 – 5; Member Show Gifts – 7<br>HPC, Spring 2002 |
| 7" TWH Mare | Black Overo Pinto (Pearled) | 684M-01 | 5 | HPC, Summer 2001 |
| 7" TWH Foal<br>7" TWH Foal<br>7" TWH Foal<br>7" TWH Foal | Black (Pearled)<br>Palomino Pinto<br>Gray Appaloosa<br>Blue Roan | 684F-01<br>684F-PP<br>684F-GA<br>684F-BR | 5<br>3<br>3<br>3 | HPC, Fall 2001<br>HPC, Spring 2004<br>HPC, Spring 2005<br>HPC, Winter 2006 |
| 7" TWH Head Lapel Pin | Lapel pin: head of 7" TWH Stallion, chestnut/ bay colors | No # | 10? | Gift to invitees to California event, 2002 |
| Tinymites | Tri-Color ("Holiday") 3 pc set: one each of Green, White, Maroon | TM-33 | 6 sets | 2 HP contests, Winter 2004, 3 winners each |
| Tinymites | Bright Gold ("Golden Ornament") 5 pc set | TM-5G | 3 sets | HPC, Fall 2005 |
| Tinymites | Chocolate Sprinkles ("Non Pareils") 5 pc set: white w/ brown flecks and brown w/ white flecks | TM-5BW | 3 sets | HPC, Spring/Summer/ Fall 2006 |

**Gift Runs** – The table above lists 19 gift model runs created as Hoof Pick Contest prizes and four other gift runs. The "Item #" is the catalog number, and usually does not appear on the model itself. HPC = Hoof Pick Contest (under "Occasion"). Seasonal dates, like "Fall 2002," are the cover date of the company newsletter issue in which the model winners were announced. "Winter" issues were for the first quarter of the year.

# Chapter 12
# Gift Editions
## 9" and 11" Series

This chapter includes 11 gift editions: two runs for Jamboree helpers (a 9" Mustang and an 11" Arabian); a Mustang run used as show trophies; and eight editions of Hoof Pick Contest prizes: 9" Thoroughbreds (two runs), 9" Five-Gaiters (two runs), and 9" Polo Ponies (four runs).

The 11 gift editions in the 9" and 11" series are:

## A & B

- Bay (Glowing) 9" Thoroughbred, 873-BG; Quantity: 3

This model is a clear, light bay with a slight glow, a star painted on its face, and hind socks with pink feet under them. It was the prize awarded to three Hoof Pick Contest winners named in the Spring 2003 company newsletter. The 9" Thoroughbred mold was not used for production models after 2000, and was only used for some test pieces and two gift runs totaling eight pieces. *Photos A & B, courtesy of Victoria Zanutto.* (Also see photo in Introduction to Part 2.)

## C

- Dapple Grey (Pearled) 9" Five-Gaiter w/ Black Mane and Tail, 881-08; Quantity: 5

This model is like "Snow Angel," the 2002 Holiday Horse, but with black mane and tail. It has a pearled grey body; four white socks; pink feet; and black knees, hocks, and muzzle. It may have been made by adding black to leftover "Snow Angels." Five were awarded to Hoof Pick Contest winners announced in the Fall 2002 company newsletter. *Photo C, courtesy of Elaine Boardway.*

## D

- Black (Dappled and Pearled) 9" Five-Gaiter, 881-BP; Quantity: 3

A pearled finish and subtle dappling enhance this black Saddlebred with four white socks and pink hooves. The company called the color: black pearl dapple. This model was awarded to the three Hoof Pick Contest winners named in the Summer 2005 company newsletter.

Robyn Porter, whose mother, Dee Gwilt, owns one, said that the entire model is pearled, including the mane and tail, socks and hooves! The eyes are tri-color with "pink/white corners, brown second, and then a black center," she said. *Photo D: model, courtesy of Dee Gwilt; photo, courtesy of Robyn Porter.*

**A**
Bay (Glowing) 9" Thoroughbred, Gift Run 873-BG.

**B**
The Bay (Glowing) 9" Thoroughbred, Gift Run 873-BG, has a painted-on star.

**C**
Dapple Grey (Pearled) 9" Five-Gaiter w/ Black Mane and Tail, Gift Run 881-08.

**D**
Black (Dappled and Pearled) 9" Five-Gaiter, Gift Run 881-BP.

## E

- Metallic Gold Bay 9" Polo Pony with Red Wraps, 883-GR; Quantity: 3

With a metallic gold body and black points, this model looks like "Heart of Gold," the 2004 Holiday (production) model, but with red wraps. Three were awarded to Hoof Pick Contest winners announced in the Summer 2004 company newsletter. *Photo E, courtesy of Victoria Zanutto.*

## F

- Metallic Gold Bay 9" Polo Pony with Blue Wraps, 883-GB; Quantity: 3

This model is metallic gold with black points, like "Heart of Gold," the production model (2004), but has medium blue, instead of white, bandages. Three like this were awarded to Hoof Pick Contest winners named in the Fall 2004 company newsletter. *Photo F, courtesy of B & K.*

**E**
Metallic Gold Bay 9" Polo Pony with Red Wraps, Gift Run 883-GR.

**F**
Metallic Gold Bay 9" Polo Pony with Blue Wraps, Gift Run 883-GB.

## G & H

- Bay Overo Pinto Polo Pony, 883-BP; Quantity: 6

This model has an apron face, pink muzzle, and bay body color with overo pinto markings painted on the neck, barrel, and hindquarters. Its pattern is about identical to "Black Tie," the black overo pinto production Polo Pony. The bay pinto was awarded to three Hoof Pick Contest winners announced in the Winter 2003 company newsletter, and three more were gifts to member shows. Elaine Boardway's show was one of them; Elaine said she used it as the ribbon buy-back raffle prize. Collector Sylvia Hand, who owns the model illustrated, said that nothing was written on the bottom of it. *Photos G & H, courtesy of Sylvia Hand.*

## I, J, & S

- Buckskin Pinto 9" Mustang with White Tail, 886-03W; Quantity: 29; VG: $30... EX: $ 40... NM: $50.

The white-tailed version of the Buckskin Pinto Mustang was a gift for 2001 Jamboree judges and volunteers, June 2001. After the Jamboree, the tail on the remaining Buckskin Pinto Mustangs was painted black. The black-tailed version was then sold to winners of a drawing open to the company's 2001 club members.

(The company initially said that there were 29 with white tail and 21 with black tail. Then, for the model list in the newsletters, the company reversed itself. Given that the list appeared more than a year after the events and included at least a few errors, I tend to think the company's first account is more likely to be correct. In any event, between 20 and 30 were made of each version.)

The owner of this model, Melanie Teller, said she got it second-hand for $50 in around 2006. *Photos I, J, & S, courtesy of Melanie Teller.*

## K & T

- Cinnamon Dapple Bay 9" Polo Pony, 883-CDB; Quantity: 3

This fantasy color is like the "Sugarplum" 2001 Holiday Horse, but with black points. The photos make him look more golden than "Sugarplum" though. The dappling is subtle. Three of the Cinnamon Dapple Bay were awarded to Hoof Pick Contest winners announced in the Winter 2005 company newsletter. Handwritten on the belly of this particular example is: 04 Hoofpick 1 of 3. In photo T, note the eyeliner and tri-color eyes. *Photos K & T, courtesy of Melanie Teller.*

**G**
Bay Overo Pinto Polo Pony, Gift Run 883-BP.

**H**
Bay Overo Pinto Polo Pony gift run – *left side.*

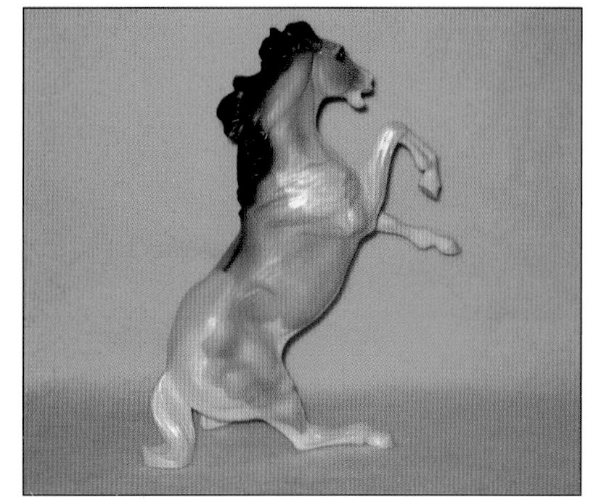

**I**
Buckskin Pinto 9" Mustang with White Tail, 886-03W.

# L

- Silver-Gray 9" Thoroughbred w/ Black Points, 873-SG; Quantity: 5

This Thoroughbred with black points and no white markings and no leg bars was awarded to Melissa Clegg and four other Hoof Pick Contest winners announced in the Summer 2002 company newsletter. Another winner sold hers on eBay December 12, 2010 for $127.50. Eight people bid.

Winning bidder Eleanor Harvey said, "His coat is a pearled, smoky silver grey with a slight iridescent look to it. There are faint pink/purple tones in the right light. It's not really metallic, but all of the pearled models in grey and brown [from Hartland 2000] have a metallic element in the paint." She called him "grulla." He has no dorsal stripe, but grulla is the real-horse color he is most like.

The company did not give the model a catalog number, so I gave it the 1960s mold number, 873, plus "SG" for "Silver Gray." *Photo L, courtesy of Melissa Clegg.*

**J**
White-Tail Buckskin Pinto 9" Mustang gift run – *left side.*

**K**
Cinnamon Dapple Bay 9" Polo Pony, Gift Run 883-CDB.

**L**
Silver-Gray 9" Thoroughbred w/ Black Points, Gift Run 873-SG.

## M, N, & O

- Black Pinto 11" series Arabian with No Star, 901-10B; Quantity: 35; VG: $30... EX: $ 40...$NM: $50.

This black tobiano pinto with black forehead (no white star painted on) was a gift from Hartland to 35 volunteers at the Jamboree in August 2002. The 35 models came out of a total run of 50 Black Pinto Arabians. After the Jamboree, a star was painted on the forehead of the 15 leftover models, which were then sold. (See Chapter 5.) So, the 15 with a star added count as a production run, and the 35 with no star were a gift run. Other features of this model are four high white stockings with pink feet under them, and white attractively spread over the withers, upper barrel, and back. He's an Arab in formal wear. *Photos M, N, & O, courtesy of B & K.*

## P, Q, & R

- Bay Pinto (Pearled) 9" woodcut Mustang; Quantity: 3

This model is chocolate brown and pearl white. Its tobiano pinto pattern is similar to the pattern on the production-run and gift-run pinto Mustangs in 2001. The mane and tail are solid black; white stockings extend above the knees and hocks; the feet and muzzle are pink. The eyes are black with a white "C" in the back corner and a white highlight.

The original owner of this model, Melanie Teller, said that it was awarded for a Hartland champion or reserve; she thinks at the Kitty's No Snow Model Horse Show in California in 2002 or 2003. Written on the bottom of the tail is: 2 of 3 2002. In May 2011, Hartland's horse director for the 2000-2007 models, said that the bay pinto woodcut Mustang was "a run of three to be used for awards."

In all the years Hartland Mustangs have been made, there's never been a bay pinto before. *Photos P, Q, & R: model, courtesy of Robert Ezerski; photos by the author.*

**M**
Black Pinto 11" series Arabian with No Star, Gift Run 901-10B.

**N**
Black Pinto 11" series Arabian with No Star gift run – *left side.*

**O**
The Black Pinto 11" series Arabian, Gift Run 901-10B, has a solid black face with no white markings.

**P**
The Bay Pinto (Pearled) 9" woodcut Mustang was a Gift (Trophy) Run.

**Q**
The white markings on the Bay Pinto (Pearled) 9" woodcut Mustang are very reflective.

**R**
The Bay Pinto (Pearled) 9" woodcut Mustang gift (trophy) run does not have a white spot on either hip.

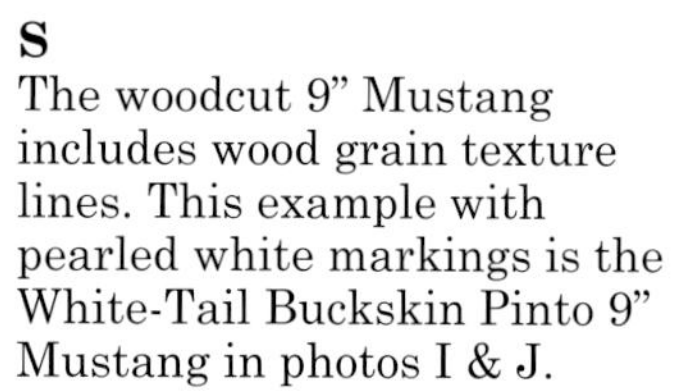

**S**
The woodcut 9" Mustang includes wood grain texture lines. This example with pearled white markings is the White-Tail Buckskin Pinto 9" Mustang in photos I & J.

**T**
The 9" series Polo Pony has a short muzzle with ears turned halfway back. This is the Cinnamon Dapple Bay Polo Pony in photo K.

# Chapter 13
# Gift Editions
## 7" and Tinymites

Members of the 7" Tennessee Walker Family, abbreviated as "TWH," were used for eight gift editions. There were three runs of stallions, four runs of foals, and one run of mares. Along with three editions of Tinymite sets and a group of horse head pins, this chapter includes a total of 12 gift editions 7" or smaller.

The 7" and smaller gift runs are:

### A & B

- Black Overo Pinto (Pearled) 7" TWH Mare, 684M-01; Quantity: 5

The pearled finish is visible on the body of this mare with a right front sock over a pink hoof, apron face, pink muzzle, and white markings on the sides and neck. Her three black feet, lower legs, mane, and tail contrast with her body by being plain black, instead of pearled. She was awarded to five Hoof Pick Contest (HPC) winners whose names were announced in the Spring 2002 company newsletter. *Photos A & B, courtesy of Lynn Isenbarger.*

### C & D

- Chestnut Tobiano Pinto 7" TWH Stallion, "Dapper," 684S-05; Quantity: 12; VG: $39… EX: $52… NM: $65.

"Dapper," the chestnut pinto, was awarded to five Hoof Pick Contest winners named in the Winter 2002 company newsletter. Also, seven of "Dapper" were gifts to member shows to be used as door prizes, show trophies, or a raffle item. One was raffled at Northwest Congress in Washington state, May 18-19, 2002.

**A**
Black Overo Pinto (Pearled) 7" Tennessee Walker Mare, Gift Run 684M-01.

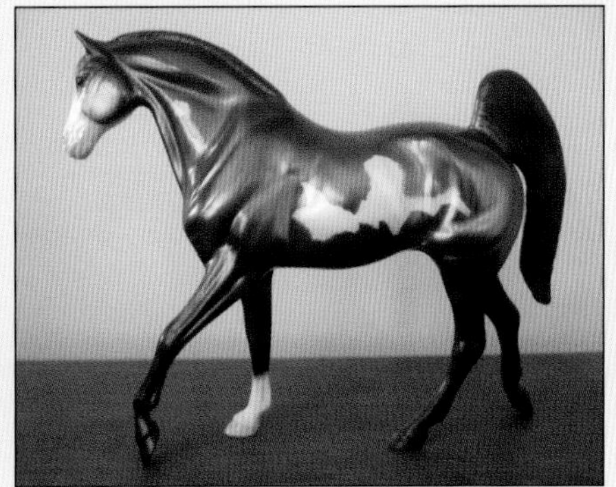

**B**
Black Overo Pinto (Pearled) 7" Tennessee Walker Mare gift run – *left side.*

**C**
"Dapper," the Chestnut Tobiano Pinto 7" Tennessee Walker Stallion, Gift Run 684S-05, has a subtly pearled finish.

"Dapper" is pale, orangy tan with tobiano pinto markings and a darker brown mane, tail, knees, and hocks. He has a left front stocking with pink hoof; the other feet and the muzzle are dark brown. Collector Deana Sprague said his stocking was crisply stenciled. She said the eyes are black "with only the tiny white highlight dot, no eye whites." She described his body color as "pearly peach dun." The model looks more pearled in the outdoor photo. *Photo C, courtesy of Deana Sprague. Photo D, courtesy of Robyn Porter.*

## E & F

- Pearled Sorrel ("Champagne") 7" TWH Stallion, "Ascot"; 684S-02; Quantity: 18; VG: $39... EX: $52... NM: $65.

"Ascot" was given to the five Hoof Pick Contest winners named in the Spring 2001 company newsletter; 13 more were given to member shows in the company's show producer's club to use as a door prize or show trophy.

"Ascot" is light chestnut over a pearl white undercoat, and has dark hooves and muzzle. The pearl white coat shows on most of his mane and tail and on his lower legs and upper leg insides. His color resembles the mare in the production Champagne Family. The company called him "amber champagne with light mane and tail," but champagne horses have largely pink skin, and he'd be a gold champagne, not amber champagne. He's a gorgeous, fantasy sorrel or chestnut. *Photo E, courtesy of Marla Phillips. Photo F, courtesy of Robyn Porter.*

**D**
In sunlight, "Dapper," the Chestnut Tobiano Pinto 7" Tennessee Walker Stallion gift run looks iridescent.

**E**
"Ascot," the Pearled Sorrel ("Champagne") 7" Tennessee Walker Stallion, Gift Run 684S-02, is the color of a mare in a production TWH family.

**F**
"Ascot," the Pearled Sorrel ("Champagne") 7" Tennessee Walker Stallion gift run, glows in the sunlight.

## G

- Palomino Pinto 7" series TWH Foal, 684F-PP; Quantity: 3

This pale, palomino pinto foal with pink feet and a light brown muzzle was awarded to the three Hoof Pick Contest winners named in the Spring 2004 company newsletter. *Photo G, courtesy of Sylvia Hand.*

## H

- Black (Pearled) 7" series TWH Foal 684F-01; Quantity: 5

Five, pearled black Tennessee Walker Foals were produced and awarded to Hoof Pick contest winners announced in the company's Fall 2001 newsletter. This foal has four white socks and peach hooves. *Photo H: model, courtesy of Dee Gwilt; photo, courtesy of Robyn Porter.*

## I & J

- Gray Appaloosa 7" series TWH Foal, 684F-GA; Quantity: 3

This appaloosa foal has a gray body with light dapples in the gray areas, black spots on a white hip blanket, four white socks, and pink hooves. He or she was awarded to three Hoof Pick Contest winners whose names were announced in the Spring 2005 company newsletter. *Photo I, courtesy of Sylvia Hand. Photo J, courtesy of Robyn Porter.*

**G**
Palomino Pinto 7" series Tennessee Walker Foal, Gift Run 684F-PP.

**H**
Black (Pearled) 7" series Tennessee Walker Foal, Gift Run 684F-01.

**I**
Gray Appaloosa 7" series Tennessee Walker Foal, Gift Run 684F-GA.

**J**
Gray Appaloosa 7" series Tennessee Walker Foal gift run – *left side.*

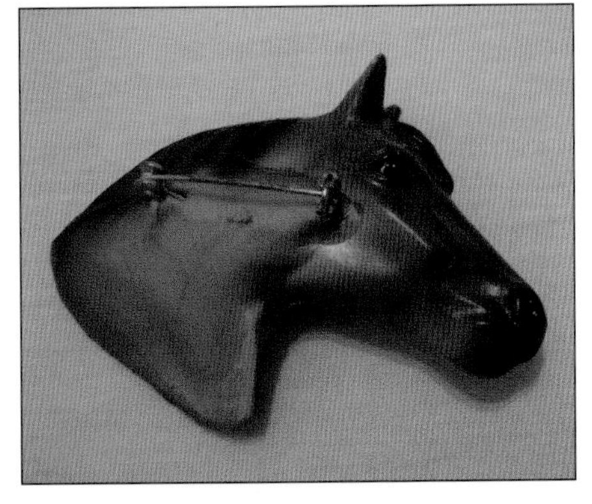

**K**
This lapel pin is the head of a 7" Tennessee Walker Stallion, painted bay.

**L**
The 7" Tennessee Walker Stallion lapel pins were a gift item. This one's bay; reportedly, others were chestnut.

## K & L

- Lapel Pin: Head of 7" TWH Stallion; Quantity: about 10

This bay, horse head lapel pin is about 2⅛" high and 2¾" wide. It depicts, and was made from, the head of a 7" Tennessee Walker stallion. Collector Marla Phillips got it second-hand from artist Kristina Francis, who said that Hartland's horse line director, Sheryl Leisure, painted the pins and gave them to about 10 people who met with her, by invitation, at Equine Affaire about 2002 [probably Feb. 9-10, 2002] to brainstorm about Hartland horses. Kristina Francis recalled "chestnuts with differing markings and eye whites," but this one is bay. *Photos K & L, courtesy of Marla Phillips.*

**M**
Tri-Color Tinymites Gift Run – Sylvia Hand received a green Arabian, pearl white Morgan, and maroon Tennessee Walker.

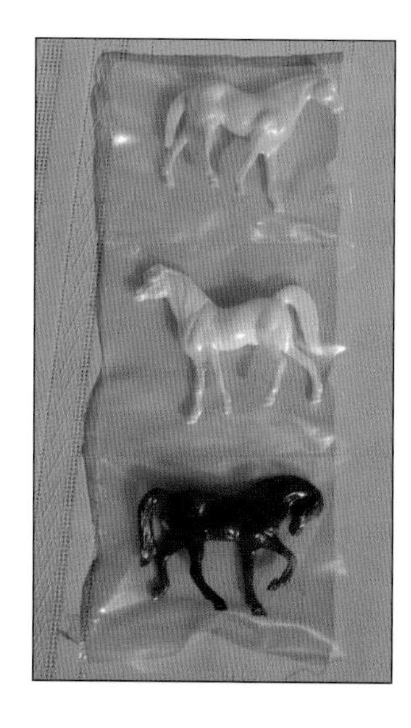

**N**
Robyn Porter's set has a green Quarter Horse, white Arabian, and maroon Morgan.

**O**
Tammy Nguyen received a green Tennessee Walker, white Thoroughbred, and maroon Arabian.

**P**
Bay Pinto 7" Tennessee Walker Stallion, Gift Run 684S-04.

**Q**
The Bay Pinto 7" Tennessee Walker Stallion has thoroughly black points.

**R**
A white Quarter Horse, maroon Arabian, and green Thoroughbred went to Marla Phillips.

**S**
Melanie Teller's set has a maroon Thoroughbred, green Quarter Horse, and white Arabian that looks creamy white.

## M, N, O; and R & S

- Tri-Color "Holiday" Tinymite 3 pc Set; Quantity: 6 sets

The company called this the "Mint Green / Maroon / Pearl White 3 pc Holiday Set," and produced six sets. They were awarded to three winners in each of two Hoof Pick Contests. All six winners were named in the Winter 2004 company newsletter.

Each set includes three different breeds – one in green, one in white, and one in maroon. The breeds are drawn from the five light horse breeds of Tinymites – there are no draft horses – but which breed is which color varied. Five of the six sets are illustrated here, and each one is different!

Each set came heat-sealed in a partitioned plastic bag. After Marla Phillips took hers out of the package, she inspected them more closely. She said: "The Tinymites are definitely painted. The red one looks kind of glazed. The white one does have a pearled finish. The green one was hard to tell, so I scratched the bottom of his hoof, and it was white underneath." (That took daring!)

The five sets illustrated include 13 different models (breed and color combinations). The two variations not shown are: a maroon Quarter Horse and green Morgan. *Photos, courtesy of Sylvia Hand (M); Robyn Porter (N); Tammy Nguyen (models in O) and Robyn Porter (photo O); Marla Phillips (R); and Melanie Teller (S).*

## P & Q

- Bay Pinto 7" TWH Stallion, 684S-04; Quantity: 5

This bay pinto stallion went to five Hoof Pick Contest winners announced in the Spring 2002 company newsletter. He has a left front stocking with pale hoof; otherwise, black points and hooves. His white, tobiano pinto markings are almost the same as on the production run Black Pinto 7" TWH stallion (684S-01). *Photos P & Q, courtesy of Doris Hunter.*

## T

- Chocolate Sprinkles ("Non Pareils") Tinymite 5 pc Set; Quantity: 3 sets

This set consists of five breeds, some white with brown flecks, and some dark brown with white flecks. Three sets were awarded to Hoof Pick Contest winners named in the final company newsletter, the combined Spring/Summer/Fall 2006 issue.

Sylvia Hand received a set with a dark Arabian and Tennessee Walker. Her set matches the illustration in the company newsletter, so it's possible that all three sets were alike. *Photo T, courtesy of Sylvia Hand.*

**T**
The Chocolate Sprinkles ("Non Pariels") Gift Run Tinymite 5-pc Set includes white horses with brown specks and dark brown horses with white specks.

## U

- Bright Gold ("Golden Ornament") Tinymite 5 pc Set; Quantity: 3 sets

The "Golden Ornament" set of bright gold Tinymites was awarded to three Hoof Pick Contest winners named in the Fall 2005 company newsletter. The horses really can be used as ornaments because each one has a metal hanging loop (maybe an eye screw) attached to it. One of the winners, Denise Brubaker, said, "The 'implementation' was done extremely neatly. There's no excess material around it, and they're all consistently lined up with the spine, so to speak." Each horse has a tag reading, "Hartland Horses 2005" on a gold cord tied through the metal loop. Denise reported that nothing is written on the underside of the tags or on the belly or hoof bottoms of the horses. *Photo U, courtesy of Denise Brubaker.*

## V

- Blue Roan 7" series TWH Foal, 684F-BR; Quantity: 3

This foal with pale, grayish body and black head and points was given to three Hoof Pick Contest winners announced in the Winter 2006 company newsletter, which arrived in May 2006. There are no white markings. *Photo V, courtesy of Melanie Teller.*

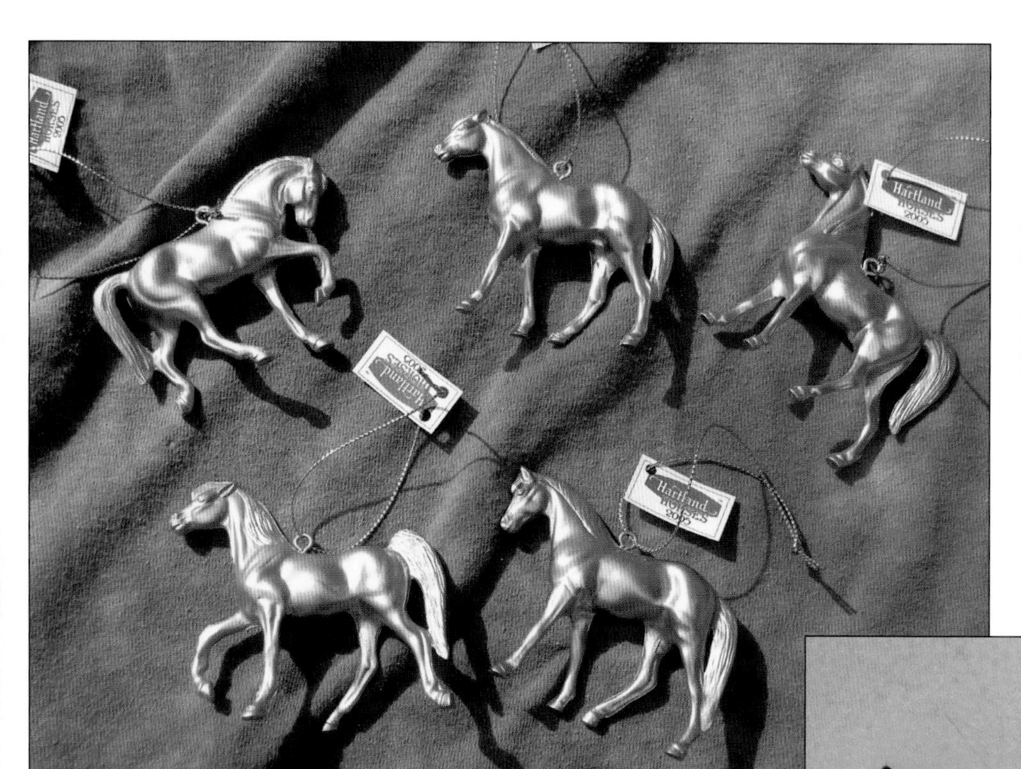


**U**
The Bright Gold ("Golden Ornament") Gift Run Tinymite 5 pc Set can be hung as decorations.

**V**
The Blue Roan 7" series Tennessee Walker Foal, Gift Run 684F-BR, has a grayish white body and black head and points.

# PART 3
# TEST (UNIQUE) MODELS

Test models, as the Hartland company used the term since 2000, were uniquely painted (one-of-a-kind) models. Some of them are similar to a production run or to another test model, but when that is the case, it seems that there is always some intentionally painted difference. For example, a test model might have white stockings or a pearled or dappled finish, but the production run similar to it does not or vice-versa.

The underside of a test model typically has a six-digit number handwritten on it, and to that, usually adds the letter "T" or the word "Test." The number identifies the model's mold number and color number, separated by a hyphen. The numbering that identifies a test model will often look something like this: 999-012T.

If instead, there is a shorter number – three to five digits – handwritten on the underside, the model is a production model from a numbered edition. Edition numbering appears, for example, as: #033 or 33/100. Models that are part of a numbered edition are obviously not test models since test models are unique.

As the name implies, test models were painted to try out a color for a possible production run. A few of the unique models were actually too elaborate to be a feasible production color, except at an astronomical price, but through them, the artist was able to enjoy wider expression and demonstrate that Hartland horses could wear a complex paint job as beautifully as any artist resin. (Some unique models were specifically painted to be a show award, and so technically, were not "test" models, but as I use the terms in this book, "unique" models and "test" models are the same thing.)

**Test Sales.** Since earlier than 2000, other model horse manufacturers have raised thousands of dollars (sometimes, partly for charity) through large public auctions of test models. No doubt, Hartland was mindful, from the start, of test models as not only a useful step in product development, but also a profit center. Hartland held large sales of test models in 2002 and 2005, and it also sold test models individually in online auctions (on eBay) from time to time. However, some of the fanciest test colors were not sold; instead, Hartland donated them as prizes at certain model horse shows in California. Also, dozens of test models have been retained, and may inspire future production runs.

**Values of Unique Models.** Information on original prices paid is included where known, but not values. It would be presumptuous to suggest values for unique objects.

It is important to note, though, that Hartland Horses' asking prices for test models were lowered by more than half between the 2002 and 2005 test sales:

The 11" series was \$200-\$225 in 2002, but in 2005, the only 11" horse in the test sale, an Arabian, was \$95.

The 9" series was \$165-\$195 in 2002, but \$65-\$75 in 2005 for 9" adults and \$45 for a 9" series foal.

The lone 7" series stallion in 2002 was \$150, but in 2005, the lone 7" series stallion in the sale was \$45.

For some categories, price comparisons were not possible. Two 7" foals were \$75 each in 2002, but none were sold in 2005. In the 2005 sale, 7" series mares were \$20-30, but absent from the 2002 test sale.

The test models do not appear on the secondary market very often, but a Polo Pony that was \$165 in Hartland's 2002 test sale was resold on eBay for \$76 in November 2010. A 9" Five-Gaiter that was \$175 in the May 2002 test sale went for \$49 on eBay late in 2010. In a handful of reported private sales, all of the sellers took a loss over what they paid when they bought the model new from the company.

The best resale price among them was a test Polo Pony that was originally $175 in 2002; it changed hands for $150 in December 2010. These prices may reflect more about the current economy than the importance of Hartland horses. Their owners usually consider them a bargain at any price.

**Test Models Not Shown.** A list of test models not shown is divided among the next four chapters, and found at the end of each of those chapters. The sources include: the 2002 and 2005 illustrated sales lists of test models, printouts of Hartland company eBay auctions and Jamboree website pages (for show award models), Jamboree programs (for raffle or auction models), and three company mailings in 2004 and 2005 that pictured test models (and asked for feedback on them).

The sales lists reported the handwritten test model numbers found on the underside of the models, but the company's eBay auctions did not. (The company would show a photo of the underside, but the number was too small for me to read.) So, for some models, only the first three digits, which denote the mold, are given. The numbers that should come after the hyphen are a mystery for now.

**A Four-Chapter Tour.** The next four chapters illustrate a good share of the test (unique) models created by Hartland since 2000. All were painted by Sheryl Leisure unless noted otherwise.

The Bay Appaloosa with Four Socks test model *(upper)* seems to bow as hunter pleasure entries dance toward the exit at an all-Arabian horse show at Earl Warren Showgrounds, Santa Barbara, California, in September 2010. *Upper photo: model, courtesy of Tammy Nguyen; photo, courtesy of Robyn Porter.*

# CHAPTER 14
# TESTS
## THOROUGHBREDS IN PLASTIC

Hartland 2000 did not issue the 9" series Thoroughbred as production models, only as gift runs and tests. Joining test Thoroughbreds in this chapter are some of the many uniquely painted Saddlebreds and Polo Ponies, plus two Mustangs, a 9" Tennessee Walker, and a 9" series Weanling Foal.

**A**
Dapple Grey 9" series Weanling Foal, Test Model 6100-05.

**B**
Grulla or Gray 9" Thoroughbred with Leg Bars, Test Model 873-010-T.

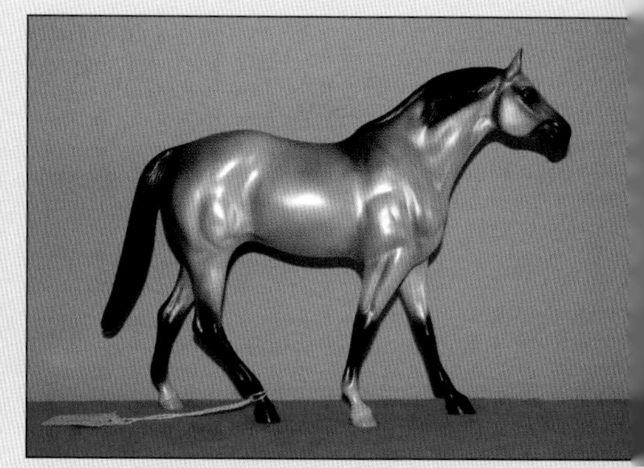

**C**
Silver-Gray (Pearled) 9" Thoroughbred with Two Socks, Test Model 873-04.

**D**
Creamy-Tan Dun Appaloosa 9" Thoroughbred, Test Model.

**E**
Palomino (Dappled) 9" Five-Gaiter, Test Model 881-016T.

**F**
Dark Sorrel 9" Thoroughbred, Test Model.

## A

- Dapple Grey Weanling Foal (9" series), Test Model 6100-05.

This foal is "matte, shaded gray with subtle white 'resist' dapple spots," according to its owner. It has white stockings, gray feet, and a darker gray muzzle, mane, and tail. (In contrast, Paola Groeber's 1988-1990 Hartland Dapple Grey Weanling production run has black feet.) The test Weanling was $45 in the December 2005 test sale. It is the only (known) painted Weanling released by Hartland 2000. *Photo A, courtesy of B & K.*

## B

- Grulla or Gray 9" Thoroughbred with Leg Bars, Test Model 873-010-T.

This model has a gray body, leg bars (horizontal stripes on the upper legs), black mane and tail, black hooves, gray lower legs, and black on the knees and hocks. A grulla would tend to have black lower legs and black on the face front. According to Sponenberg (*Equine Color Genetics*, 1996, p.55), grey horses sometimes exhibit "primitive marks" such as leg bars. This model was a reserve champion prize at the JSC show at Jamboree, October 2003. *Photo B, courtesy of Marla Phillips.*

## C

- Silver-Gray (Pearled) 9" Thoroughbred with Two Socks, Test Model 873-04.

Except for two white socks (right fore and left hind), this Thoroughbred has black points (mane, tail, and lower legs) with two black feet and pink feet under the white socks. Its finish is pearled (iridescent). (There was a gift-run Thoroughbred in silver-gray, but it has entirely black points with no white socks.) *Photo C, courtesy of Victoria Zanutto.*

## D

- Creamy-Tan Dun Appaloosa 9" Thoroughbred, Test Model.

The owner's description includes: a matte, creamy tan body; dark-brown-to-black mane and tail; and brown dorsal, shoulder, and leg stripes. There are brown spots on the white hip blanket and elsewhere on the body, and "white, dapple-like spots around the white hip blanket." The base color falls in the dun and grulla group. The owner paid $75 in 2004. Written on the bottom is: Test SL. (No doubt, "SL" stands for "Sheryl Leisure," painter of the model.) *Photo D, courtesy of B & K.*

## E

- Palomino (Dappled) 9" Five-Gaiter, Test Model 881-016T.

With subtle dapples, four white socks, and pink hooves, this medium-to-light shade of palomino was $175 in the May 2002 test sale. It was described as a test for the dappled palomino production runs of the 11" series Quarter Horse and Arabian. (The palomino production Five-Gaiter, from 2001, was a red-gold palomino and not dappled.)

In the last months of 2010, it was sold on eBay for $49.01; Eleanor Harvey, the new owner, said, "She has nice 'resist' dapples on her shoulders and on her entire left side." On the right side, the dapples are only on the shoulders. *Photo E, courtesy of Eleanor Harvey.*

## F

- Dark Sorrel 9" Thoroughbred, Test Model.

Collector Melanie Teller recalled this model as a trophy for the champion or reserve in the Hartland division at Kitty's No Snow Model Horse Show in 2002 or 2003. (The "Kitty's" show in Arcadia, California, in January 2004, listed in the North American Model Horse Shows Association online archive, may have been the last of three annual "Kitty's" shows.)

The darker areas of her body; and her mane, tail, and lower legs (except for hind socks) are very dark brown, not black, according to her owner. She has lighter (tan) areas at her muzzle, flanks, and elsewhere. Melanie wrote, "I love her mealy face." Written on her underside is "Test," and, hard to read, numbers starting with "200." With another legible digit, we would know the year. *Photo F, courtesy of Melanie Teller.*

**G**
Black Appaloosa 9" Five-Gaiter, Test Model 881-025T.

**H**
Black Appaloosa 9" Five Gaiter Test Model – *left side.*

**I**
Silver-Gray (Pearled) 9" Five-Gaiter with Four Socks, Test Model 881-10.

## G & H

- Black Appaloosa 9" Five-Gaiter, Test Model 881-025T

This test appaloosa's irregularly-shaped black spots seem to float on a white blanket that extends forward to the withers. Details include: some fine dappling in the black areas on the body and neck; four white socks; pink feet; and white (with black spots) on the face. This model was given as the trophy for the Overall Champion of the Hartland division at the "Back in Time" model horse show in Mariposa, California, hosted by Hartland 2000's horse director, Sheryl Leisure, in September 2007. *Photos G & H, courtesy of Victoria Zanutto.*

## I

- Silver-Gray (Pearled) 9" Five-Gaiter with Four Socks, Test Model 881-10

This Five-Gaiter has an iridescent, silver-gray body with four white socks and black knees, hocks, mane, tail, muzzle, and hooves. Its portrait is in Chapter 16 (photo S). *Photo I, courtesy of Victoria Zanutto.*

## J

- Blue Roan Overo-Pinto 9" Polo Pony, Test Model

This roan Polo Pony has a black head, except for an apron face and peach muzzle; black points and hooves; white leg and tail wraps; and overo pinto markings fairly similar to those on "Black Tie," the black overo pinto production model. The black on his legs goes well above his knees and hocks, almost up to the elbows and stifles. This model is marked: 883 – Test. Its owner, Victoria Zanutto, describes it as a grey roan overo pinto. Its body is gray, not blue. *Photo J, courtesy of Victoria Zanutto.*

**J**
Blue Roan Overo-Pinto 9" Polo Pony, Test Model.

**K**
Gray Semi-Leopard Appaloosa Polo Pony, Test Model.

**L**
The Gray Semi-Leopard Appaloosa Polo Pony Test Model has peach hooves on the right side and black hooves on the left.

## K & L

- Gray Semi-Leopard Appaloosa Polo Pony, Test Model

This appaloosa is white from withers to tail with a light gray forehand and legs and large and abundant leopard spots. The spots were painted freehand. It has an "arrow star" pointing upward on its forehead. That marking was first used on 1960s Hartland Polo Ponies.

The original owner, Melanie Teller, thought she first saw it in a test model display at Jamboree in 2001 or 2002. She eventually bought it, probably at the Peter Stone Company's Stone Horses Jubilee, October 8, 2005, in Pomona, California. (The Jubilee more or less took the place of the Jamboree.) *Photos K & L, courtesy of Dianne S. Teachworth.*

**Tests for "Sugarplum."** Photos M, N, and P illustrate three tests for the 2001 Holiday Polo Pony, "Sugarplum." I've dubbed the tests: Light, Medium, and Dark.

## M

- Light Test for "Sugarplum" Polo Pony, Test Model 883-017T

The lightest of three "Sugarplum" tests sold in May 2002, this one is both dappled and pearled. A golden neck sash (not shown) was originally sold with the model; it was part of the test. The catalog called it "dapple light pearl gold," and said it had pearled gold hooves, and pearl white tail, wraps, and feet. The price was $175. *Photo M, courtesy of Dianne S. Teachworth.*

**M**
The Light Test for the "Sugarplum" Polo Pony is both dappled and pearled, and has gold feet. Test Model 883-017T.

## N

- Medium Test for "Sugarplum" Polo Pony, Test Model 883-016T

This model is dappled (but not pearled), and has pink hooves and muzzle, and a white mane and tail, face, and wraps. The burgundy neck sash was part of the factory's test. In the May 2002 test sale, it was $165 and called: dapple bronze gold. On November 21, 2010, the original owner sold it on eBay for $76.03. Three people bid on it. Since a decade ago, values on collectibles have taken a hit. *Photo N: model, courtesy of Chelle Fulk; photo, courtesy of Andrew S. Culhane.*

## O

- Cream (Pearled) 9" Polo Pony, Test Model 883-006T

No production runs were similar to this pearl white model with pink shading and blue eyes. The factory called it: perlino. The May 2002 test catalog accidentally omitted its price, which was probably $165 or $175, like the other test Polo Ponies in that catalog. *Photo O, courtesy of Pamela Pramuka.*

## P

- Dark Test for "Sugarplum" Polo Pony, Test Model 883-0018T

This model is both dappled and pearled. It has light chocolate dapples; brown mane, tail, and pasterns; and gray muzzle and hooves, according to the May 2002 test sale catalog. The catalog called it "dapple chocolate pearl," and noted that it "did not have the 'cinnamon' color we were looking for."

Of the three "Sugarplum" tests, this one has the darkest feet. It also has the darkest body although that may not be apparent due to differences in the photography. The price was $175. *Photo P, courtesy of Pamela Pramuka.*

**N**
The Medium Test for the "Sugarplum" Polo Pony is dappled and has pink feet. Test model 883-016T.

**O**
Cream (Pearled) 9" Polo Pony, Test Model 883-006T.

**P**
The Dark Test for "Sugarplum" Polo Pony is dappled and pearled, and has chocolate brown feet. Number: 883-0018T.

## Q

- Brown (Dappled) 9" Five-Gaiter, Test Model 881-05

This model has a light brown body with slight shadings and subtle dapples; mostly dark mane, tail, knees, and hocks; four white socks, painted free-hand; and pink hooves. It was $75 in the 2005 test sale. *Photo Q, courtesy of Eleanor Harvey.*

**Q**
Brown (Dappled) 9" Five-Gaiter, Test Model 881-05.

## R

- Copper Chestnut 9" Five-Gaiter, Test Model 881-07

This Saddlebred in warm, copper chestnut with dark mane, tail, and hooves was $65 in the 2005 test sale. It is not dappled. No production Five-Gaiters were similar to it. *Photo R, courtesy of Eleanor Harvey.*

**R**
Copper Chestnut 9" Five-Gaiter, Test Model 881-07.

## S & T

- Black Mustang (9" series woodcut), Test Model 886-007T

This Mustang is all black, except for a crescent-shaped star. His color is normal black, with no pearling or dappling. No production run (after 2000) was made of black Mustangs. His price was $175 in the May 2002 test-color sale. Apparently, he did not sell, and was subsequently awarded as the trophy for the Reserve Grand Champion Hartland at the open model horse show at Jamboree, August 2002. *Photos S & T, courtesy of Melanie Teller.*

**S**
Black Mustang (9" series woodcut), Test Model 886-007T.

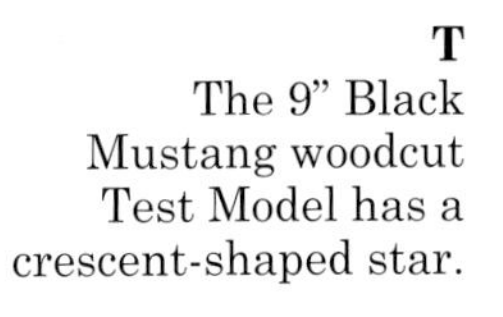

**T**
The 9" Black Mustang woodcut Test Model has a crescent-shaped star.

## U & V

- Bay Blanket Appaloosa 9" woodcut Mustang, Test Model

This bay, spotted-blanket appaloosa has black spots on his white hip blanket. He has four white socks, peach hooves, and black above the socks that extends higher than the knees and hocks. His mane, tail, and muzzle are black.

Owner Melanie Teller said she bought him from someone who won him at one of Sheryl Leisure's model shows. He's the only test model in this book known to have been painted by someone other than Sheryl Leisure. The painter was one of the partners who bought Hartland in 2000, and did not want his name published. Written on the model's belly, following his first initial and last name is: 2/10/02 Test. *Photos U & V, courtesy of Melanie Teller.*

## W

- Number 18 Rose Grey (Dappled) 9" Five-Gaiter, Test Model 881-018T

This May 2002 test-sale model is dappled with entirely solid black points and a (white) star on its face. The sale catalog described a white basecoat, light brown dapples, black shading, and a "very pink cast, too pink for production." The production rose grey, which I call Rose Grey aka Dapple Grey with Bay Shadings (February 2002, #881-03) does not have a star, has a light tail tip, and does not look pinkish! The price on this test model was $195. *Photo W, courtesy of Pamela Pramuka.*

## X

- Black (Pearled) 9" Tennessee Walker, Test Model 886-007T

This pearled, black test model has no white. In contrast, the black production model (887-02, December 2001) has white socks and a star, and is not pearled. The test was $175 in the May 2002 test sale. His body is pearled black while his points are regular black. His owner, Elaine Boardway, said that his black points "have enough of a shine that they blend with the pearl color" of his body. She said his number, written in felt tip pen on his belly, is hard to read. It's too bad those numbers seem to fade, but we know his number from the 2002 sales list. *Photo X, courtesy of Elaine Boardway.*

**U**
Bay Blanket Appaloosa 9" woodcut Mustang, Test Model.

**V**
Bay Blanket Appaloosa 9" woodcut Mustang, Test Model *– left side.*

**W**
Number 18 Rose Grey (Dappled) 9" Five-Gaiter, Test Model 881-018T.

**X**
Black (Pearled) 9" Tennessee Walker, Test Model 886-007T.

## Test Models Not Shown: 9" Thoroughbreds, Polo Ponies, and Mustangs

### 9" Series Thoroughbreds – Tests:

- Dark Rose Grey (Dappled) 9" Thoroughbred – Test 873-

A test-color 9" series Thoroughbred described as "Dark Dapple Rose Grey" auctioned on eBay on May 30, 2001. It was bid up to $255 with 40 minutes to go. Painted on November 15, 2000, it has solidly dark points and a dark muzzle. The auction stated, "The sale of these models will help fund repairs to the original Hartland horse molds."

- Flea-bit Grey 9" Thoroughbred – Test 873-02

This test is white except for a dark muzzle, dark feet, and tiny dark specks in its coat. It was $75 in the December 2005 test sale.

### 9" Mustang Woodcut – Test:

- Claybank Dun (Pearled) 9" Mustang – Test 886-009T

This model was $185 in the May 2002 test sale. It has a pale, orange-brown body with a bald face, four white socks, pink feet, and a dorsal stripe. The mane, tail, and muzzle are darker than the body. (In contrast, the1988-1990 Red Dun Mustangs have black hooves and are from the smooth Mustang mold, not the woodcut mold.)

### 9" Polo Ponies – Tests:

- Palomino (Dappled) Polo Pony – Test 883-01

This test is dappled, has a matte finish, and a masked "star blaze." The 2002 production palomino, 883-05, was pale and not dappled. The test was $65 in the December 2005 sale.

- Golden Bay Appaloosa Polo Pony – Test 883-04

This test has a bald face. It was $65 in the December 2005 test sale.

- Number 10 Red-Bay Blanket Appaloosa Polo Pony – Test 883-010T

This model is like the "Chart the Course" bay blanket appaloosa, but has a slightly more reddish body color, and has black spots on the blanket, instead of dark brown.

- Number 10 (no T) Bay Blanket Appaloosa Polo Pony – Test 883-10

Another test for "Chart the Course," this one differs from the regular run in two ways: It has black, instead of brown, hip spots, and its white blanket is longer. It extends to the withers, instead of ending just forward of the hips. The model's belly is marked 883-10, which was the catalog number of the regular run. The number handwritten on the regular run is not 883-10, but a sequence number, such as 27 or 101, since the regular run was a numbered edition.

- Number 5 Bay Blanket Appaloosa Polo Pony – Test 883-05

The white blanket on this test for "Chart the Course" extends forward to the withers, farther than on the regular run. It has brown spots on the blanket, like the regular run. It was $65 in the December 2005 test sale.

- "Yellow Gold Appaloosa" Polo Pony – Test Model 883-020

Gold leg and tail wraps make this model with golden body and black points and hooves, from the December 2005 test sale, stand out. The black spots on its hips and flanks were hand-airbrushed. It was $65.

We're not done with test Polo Ponies: Chapter 16 illustrates some more.

# CHAPTER 15
# TESTS
## A REGAL AIR

Hartland's 11" (or "Regal") series is a fine canvas for painted details. In this chapter, unique (test color) Regal Quarter Horses, Arabian Stallions, and "Lady Jewel and Jade" put their best hoof forward.

**A**
Bay Appaloosa with Hip Waders (11" QH mold), Test Model 902-T012.

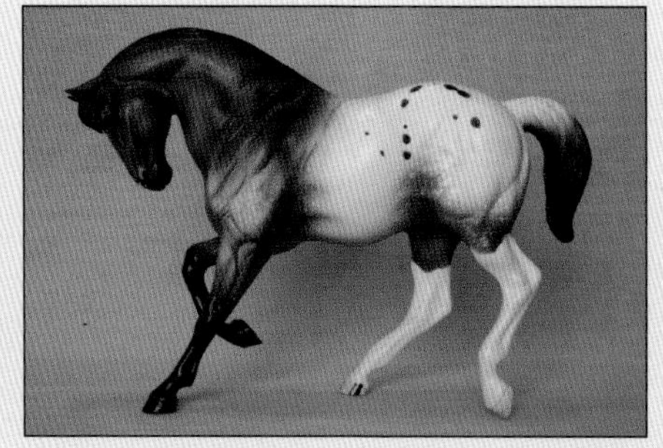

**B**
The Bay Appaloosa with Hip Waders test model – *left side.* Its spots are dark brown.

**C**
Palomino Appaloosa with Tan Spots (11" QH mold), Test Model 902-008T.

**D**
Palomino Appaloosa with Tan Spots test model – *left side.* Two of its hooves are striped.

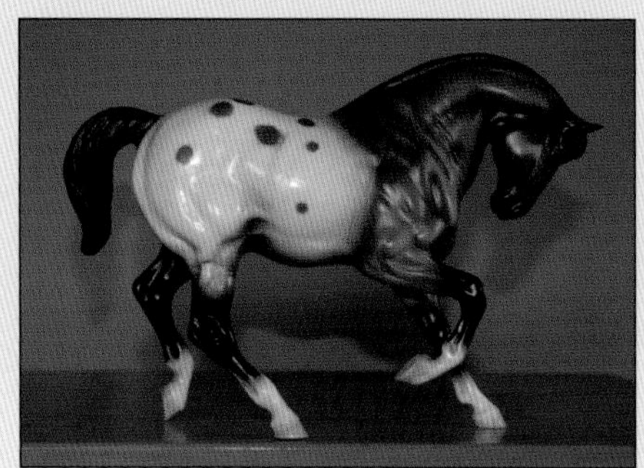

**E**
Bay Appaloosa with Four Socks (11" QH mold), Test Model 902 0005-T.

**F**
The Bay Appaloosa with Four Socks test model – *left side.* It has peach feet.

## A & B

- Bay Appaloosa with Hip Waders (11" QH mold), Test Model 902-T012

This brown-bay Appaloosa has fine dappling in the brown areas, an extended white hip blanket with brown spots, high and masked hind stockings, and stripes on the left hind hoof. The white stockings on the hind legs extend well above the hocks, like the hip waders fishermen use in streams. The front lower legs and bottom two-thirds of the tail are black. This model was awarded to the Overall Champion in the Hartland division at the JSC model horse show at Jamboree 2004. *Photos A & B, courtesy of Marla Phillips.*

## C & D

- Palomino Appaloosa with Tan Spots (11" QH mold), Test Model 902-008T

The Palomino Appaloosa with Tan Spots (11" QH mold), Test Model was the trophy for the Overall Reserve Champion Hartland at Kitty's No Snow Model Horse Show, January 11, 2003, in Arcadia, California. The first owner, Melanie Teller, called this model, "a very pretty, soft creamy palomino blanket appaloosa."

The spots on the white hip blanket are golden tan, darker than the body. It has four white socks, white mane and tail, peach hooves, and two ermine spots on the left hind pastern, and black hoof stripes below the spots. (A different palomino appaloosa test model, sold in May 2002, has spots that are larger and paler.) *Photos C & D, courtesy of Pat Noble.*

## E & F

- Bay Appaloosa with Four Socks (11" QH mold), Test Model 902 0005-T

Hartland sold this test, 11" series bay Appaloosa on eBay on September 13, 2001, for $305. It has few, but large, brown spots on the rump. The hip blanket extends forward to the withers. The model has peach hooves and four, low white socks, with black above them that goes above the knees and hocks. His belly number is somewhat unusual in its format. It is: 902 0005-T. *Photos E & F: model, courtesy of Tammy Nguyen; photo, courtesy of Robyn Porter.*

## G & H

- Dun / Grulla 11" Quarter Horse, Test Model 902-007T

This test Quarter Horse has a pale, golden tan body, black points, a dark brown head, and faint leg bars, plus a faint dorsal stripe and shoulder stripe. Hartland Horses sold him on eBay, January 28, 2002. (The bid was already $204 with four days to go!) The current owner bought him second-hand in 2005. *Photos G & H, courtesy of Marla Phillips.*

**G**
Dun / Grulla 11" Quarter Horse, Test Model 902-007T.

**H**
The Dun / Grulla 11" Quarter Horse test model – *left side.* It has faint, reddish leg bars (horizontal stripes above the knees and hocks).

## I & J

- Caramel Bay Tobiano Paint (11" QH mold), Test Model 902-001T

This early test on the Regal (11" series) Quarter Horse mold was Lot #33 in the Jamboree Benefit Auction at Jamboree, June 2001. The company called the model: caramel buckskin color. It has pink feet and the same pinto pattern used for the 2001 production black pinto, "Tom-Tom" (901-01). Written on the test's belly is: 902-001T. *Photos I & J, courtesy of Marla Phillips.*

**I**
Caramel Bay Tobiano Paint (11" QH mold), Test Model 902-001T.

**J**
Caramel Bay Tobiano Paint test model – *left side.*

## K

- Buckskin or Dun (Dappled) "Lady Jewel" mold, Test Model 904M-002T

This 11" series cantering Arabian mare has a light, drab brown body with subtle dappling in dark shadings; four white socks, pink feet, and a dorsal stripe (but no leg bars). Hartland donated her and her nearly matching foal to the Saturday evening raffle at Jamboree in June 2001, and Denise Hauck won them. They were raffled as a pair. *Photo K, courtesy of Denise Hauck.*

## L

- Buckskin or Dun "Jade" mold, Test Model 904F-002T

This frolicking, 11" series Arabian foal looks related to her mother, but is not an exact match. She has four white socks of varying height, but, unlike her dam, she has no dapples. Owner Denise Hauck said, "She does have a dorsal stripe, so I guess I would call her color 'baby dun,' but she doesn't have leg barring. She is much lighter than the mare." Along with the mare, the filly was raffled at Jamboree in June 2001. *Photo L, courtesy of Denise Hauck.*

**K**
Buckskin or Dun (Dappled) "Lady Jewel" mold, Test Model 904M-002T.

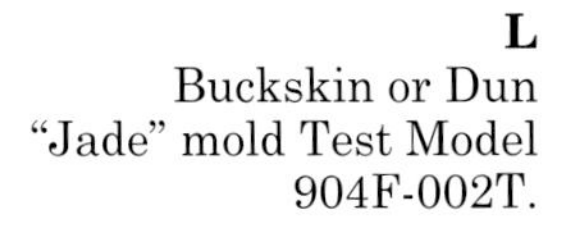

**L**
Buckskin or Dun "Jade" mold Test Model 904F-002T.

## M & N

- Light Bay (Dappled) 11" Arabian, Test Model 901-T018

This bay Arabian with four white socks and pink feet was the Overall Hartland Division award at Kitty's No Snow Show, 2005, hosted by Kitty Cantrell and Caroline Boydston. It has fine, light brown dapples in the dark-shaded areas. *Photos M & N, courtesy of Victoria Zanutto.*

**M**
The Light Bay 11" Arabian, Test Model 901-T018 has very fine light brown dapples.

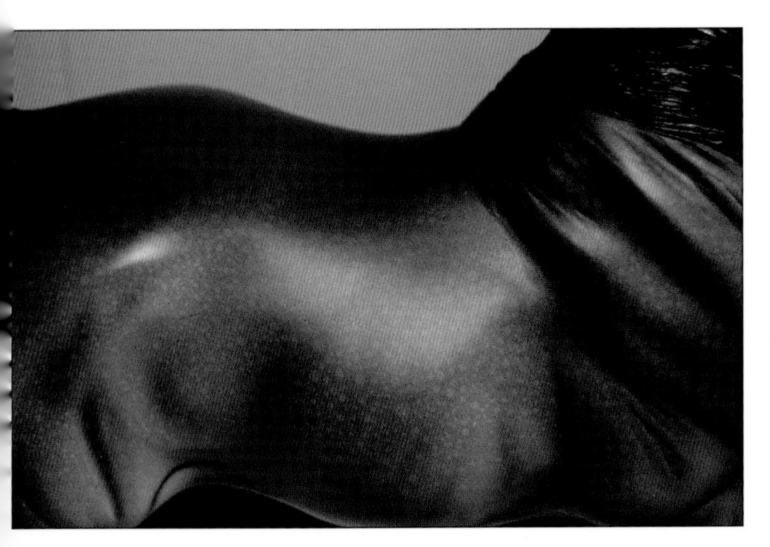

**N**
After a mist of oil was sprayed on the Light Bay 11" Arabian test model, each tiny bead of oil resisted the darker paint added later, and became a dapple.

## O, P, Q, & R

- Bay, Semi-Leopard Appaloosa 11" Arabian, Model #901-020 Test

Hartland donated this model as the Hartland Division Champion Award at the Jamboree Winter Challenge model horse show, February 28-March 1, 2009, in Mariposa, California. His owner, Melanie Teller, won him. He's a semi-leopard bay appaloosa with brown spots on his hindquarters, barrel, shoulders, and neck. He has fine speckling on the muzzle, and striping visible on three of his hooves. His tail is black except for white guard hairs. His mane is black, and he has black on his knees and hocks above four white stockings. *Photos O, P, Q, & R: courtesy of Melanie Teller.*

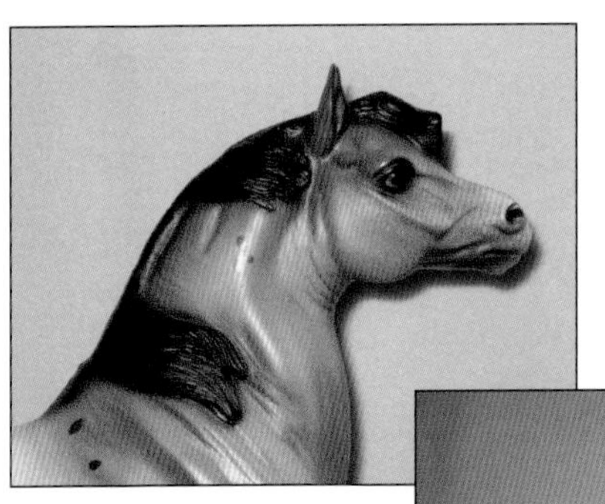

**O**
The chin of the 11" series ("Regal") Arabian is raised, and he has a heavy forelock. This is the Bay, Semi-Leopard Appaloosa Test Model 901-020.

**P**
The Bay, Semi-Leopard Appaloosa 11" Arabian, Test Model 901-020, has a complicated and realistic pattern of spots.

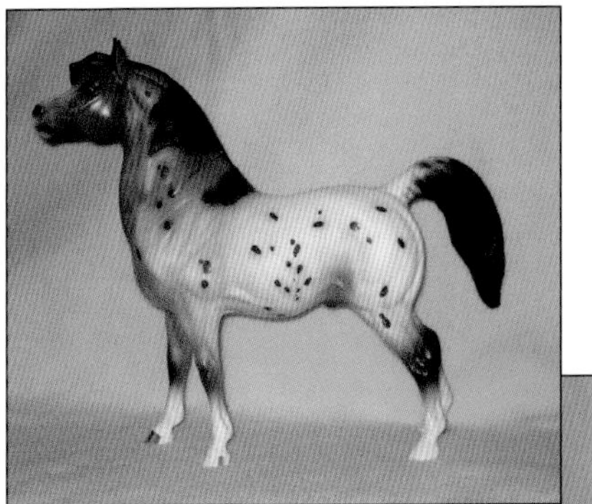

**Q**
The white guard hairs at the top of the tail are an interesting touch on the Bay, Semi-Leopard Appaloosa 11" Arabian test model.

**R**
The Bay, Semi-Leopard Appaloosa 11" Arabian test model has fine speckling on his muzzle and face.

## S, T, U, V, W, & X

- Chestnut 11" Arabian (Two Colors in One), Test Model 901-010T

This Arabian is two chestnut colors in one: Light Chestnut with flaxen mane and tail on the right side, and Copper Chestnut with darker brown mane and tail on the left side. The light chestnut color was not followed by a production run, but Hartland did make a Copper Chestnut 11" Arabian production run (901-04) in July 2001. (The production models have only two white socks, in front, while this test has four white socks.)

In the portraits, you can see that his forelock is light on one side and dark on the other. The model has no dapples, but the Light Chestnut side looks pearled.

This "horse of another color" was originally offered as a silent auction item with a minimum bid of $225 in the May 2002 test sale. However, owner Melanie Teller bought him from Hartland Horses on eBay for $100 in March of 2003. She called the eBay price "a steal," and said that the model has repeatedly qualified for the North American Nationals model horse show in the collectability division. *Photos S, T, U, V, W, & X: courtesy of Melanie Teller.*

## Test Models Not Shown
## 11" Quarter Horses, Arabian Stallions, and "Lady Jewel and Jade"

## 11" Series Quarter Horse Mold – Tests:

- Palomino Appaloosa (11" QH mold) – Test 902-004T

In the May 2002 test sale, this palomino blanket appaloosa had a minimum asking price of $200. It has a medium golden-yellow body color with "rusty body shading," four socks, pink feet, and dapples where the white blanket meets the golden body color. The flyer said it has "a long blaze with a pink muzzle," but from the side view, the end of the muzzle is dark gray or black. The spots on the blanket are relatively large and match the golden-yellow body color. (A different test palomino appaloosa on the QH mold has a paler body and smaller and darker spots, among other differences.)

This model may not have sold in May 2002 because I believe it was pictured in the company's Winter 2003 newsletter (in black-and-white). The newsletter said the model had recently been sold on eBay.

- Buckskin Appaloosa (11" QH mold) – Test 902-006T

In the May 2002 test sale, a buckskin blanket appaloosa was a silent auction item with a minimum bid of $200. It has a long blaze with pink muzzle, ermine spots and hoof stripes, three socks (all but the right fore), and "dapples where the blanket meets the body color." It was a test for the September 2001 production run (902-03), which has only one white sock and has no dappling.

A collector recalled a test model Buckskin Appaloosa (11" QH mold) being awarded to the Overall Grand Champion Hartland at the Jamboree's JSC (open) show in August 2002, but confirmation of that was unavailable. If the model in the May 2002 test sale did not sell, it could have been a show prize.

- Shaded Bay (?) 11" Quarter Horse – Test 902-

A prize for a champion at the JSC model show at Jamboree 2003 was an model from the 11" QH mold that appears, in a small photo printed in black-and-white from the Jamboree website, to be a shaded bay. It has a left hind stocking and a white marking on its face that, in the right profile, ends with a narrow white peninsula in mid-cheek, at eye level.

- Dun Overo Paint (11" QH mold) – Test 902-

The 11" QH mold in pale, claybank dun with a frame overo pinto pattern was Lot #17 in the Jamboree Benefit Auction at Jamboree in October 2004. It has very high white stockings in front – well past the knees – and white on the face.

Hartland donated the model to the auction. The minimum bid was $35; I do not know the winning bid. The number written on the underside of this unique model was not reported in the Jamboree's printed program.

S
The 11" Arabian has texture lines near the throat. This is Chestnut Arabian Test Model 901-010T. He's two colors in one.

V
The two-sided Chestnut Arabian test model, Copper Chestnut side, has an alert, but gentle eye.

T
The Light Chestnut side of the Arabian Test Model 901-010T has a flaxen mane and tail.

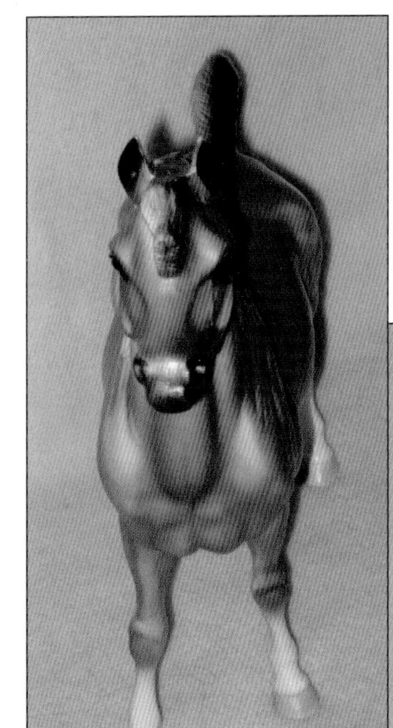

W
The color divides down the center of the Chestnut Arabian Test Model.

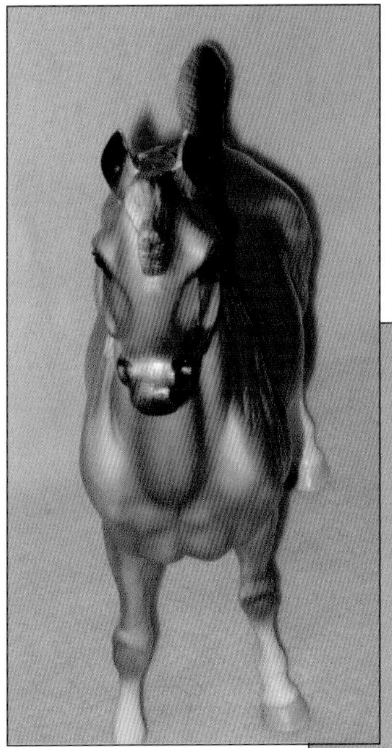

X
The left side of his tail and mane are much darker than the right.

U
The mane and tail on the Copper Chestnut side of the Arabian Test Model 901-010T are mainly darker brown than the body.

## 11" Series Arabian Stallions – Tests:

- Buckskin (Dappled) 11" Arabian – Test 901-

This test model in "dapple buckskin" was the prize awarded to the Overall Champion in the Factory Custom division at the Jamboree Summer Challenge("JSC") model horse show during Jamboree in June 2001.

- Gray (Dappled and Pearled) 11" Arabian – Test 901-

This model described as "test dapple gray pearl" was awarded to the owner of the Reserve Champion in the Factory Custom division at the JSC model show during Jamboree in June 2001. In contrast, a production run ("special run") sold at Jamboree in 2001 had a silver-gray-colored body and was pearled, but not dappled.

- Flea-bit Grey (Pearled) 11" Arabian – Test 901-006T

This gray-bodied test model with black points, except for white socks in front, has a pearled finish and tiny, dark specks all over the body, and some on the front hooves. The front hooves are (mostly) pink. The company called it "flea-bit" grey although that term normally denotes a grey horse with brown or reddish, rather than black or gray, specks. The regular run (901-02, from June 2001) is similar, but does not have the specks. The test was sold in May 2002; the minimum asking price was $200.

- Copper Chestnut (Dappled) 11" Arabian – Test 901-007T

A May 2002 test-sale model with a minimum asking price of $200, this piece is similar to the regular run Copper Chestnut (901-04, from July 2001), but has dapples whereas the regular run does not. Also, the test has four white socks while the production model has socks only in front.

- Black (Dappled and Pearled) 11" Arabian – Test 901-009T

In the May 2002 test sale, this all-black test model had a minimum asking price of $200. Then, it appeared in the December 2005 test sale with a price of $95. The production run (902-03, from June 2001) does not have dapples and has four white socks.

- Dark Bay 11" Arabian – Test 901-

A test color, 11" series Arabian in "dark, shaded bay with four white socks and pink hooves" and tri-colored eyes sold on eBay on May 11, 2004. With two hours to go, the high bid was $30.

## Lady Jewel and Jade Test Sets:

Only test sets, no production runs or gift runs, have been made since 2000 of "Lady Jewel and Jade," the 11" series cantering Arabian mare and foal. The belly numbering for each of these models would begin with 904-.

- Light Chestnut "Lady Jewel and Jade" – Test Set (Jamboree 2003)

I know of two, light chestnut "Lady Jewel and Jade" test color sets. They are not the same. The "Lady Jewel" in Lot #28 of the 2003 Jamboree benefit auction (October 24-26, 2003) has high stockings on both left legs. The stockings look masked. Her left hind hoof has a dark, vertical stripe as seen in the black-and-white photo in the Jamboree program. "Jade," the foal, has four white stockings, and his hooves look black.

- Light Chestnut "Lady Jewel and Jade" – Test Set (Jamboree 2004)

This "Lady Jewel and Jade" set was donated by Hartland Horses as a prize awarded to one of the champions at the JSC model horse show during Jamboree, October 9-10, 2004, in Pomona, California.

This set was pictured in color at the Jamboree website. The color, especially on the foal, was softly freehand-airbrushed. The mane and tail on both models are a lighter, golden-brown than the body, but are not distinctly lighter because they were not masked off. The hooves on the foal are gray, instead of black. The mare does not have a hind white stocking. Those are key features distinguishing this set from the 2003 Auction set.

More 11" test Arabian Stallions are illustrated in Chapter 17.

## CHAPTER 16

# TESTS

## PINTO PLEASURE

Hartland's Tennessee Walkers look like a pleasure to ride, and Hartland 2000 painted some interesting pinto markings. This chapter includes more unique ("test color") models painted since (or during) 2000.

**A**
Chestnut 9" Tennessee Walker, Test Model 887-00ST.

**B**
Chestnut 9" Tennessee Walker test model – *left side.* All of the 9" Tennessee Walkers are woodcuts (have a whittled texture).

**C**
Bay Pinto 7" Tennessee Walker Foal, Test Model 684F-009.

**D**
Chestnut Pinto 7" Tennessee Walker Foal, Test Model 684F-010.

**E**
Chestnut Pinto 7" Tennessee Walker Stallion, Test Model 684S-011T.

**F**
Chestnut Pinto 7" Tennessee Walker Stallion test model – *left side.* In this outdoor photo, his color looks brighter.

## A & B

- Chestnut 9" Tennessee Walker, Test Model 887-00ST

The 9" Tennessee Walker depicts relaxed and fluid action. This warm chestnut was awarded to the Overall Hartland Reserve Champion at the JSC model show at Jamboree in Pomona, California, in October 2004. It has a lighter mane and tail, dark front hooves, pink hind hooves, and hind (white) socks. Owner Victoria Zanutto, who won it, wrote, "It's really very pretty." His close-up is in photo T. *Photos A & B courtesy of Victoria Zanutto.*

## C

- Bay Pinto 7" Tennessee Walker Foal, Test Model 684F-009

This foal has an orange-tan body color with four white socks, star face, pink hooves, and masked, tobiano pinto markings, plus black on the knees, hocks, tail, and upper mane. Its pinto pattern was used on two production runs of Tennessee Walker Foals: Black Pinto (684F-03) and Bay Pinto (684F-02). However, the production bay pinto has a light, dusty brown, rather than orange-tan, body color. In the May 2002 test sale, the bay pinto was called: red bay tobiano pinto. The price was $75. *Photo C, courtesy of Sylvia Hand.*

## D

- Chestnut Pinto 7" Tennessee Walker Foal, Test Model 684F-010

Like the bay pinto test foal, this model has an orange-tan body color with four white socks, star face, and pink hooves. Its pattern of masked, tobiano pinto markings was later used for the production run foals in black pinto and bay pinto, but no production run was made in chestnut pinto. The May 2002 test sale listed the foal as: red chestnut tobiano pinto. ("Bright" would be a better descriptor than "red.") In the company's model number, "F" stands for "foal." It was $75. *Photo D, courtesy of Sylvia Hand.*

## E & F

- Chestnut Pinto 7" Tennessee Walker Stallion, Test Model 684S-011T

This was a test for "Dapper," the gift run chestnut pinto 7" TWH stallion. According to his owner, Deana Sprague, the test's body is "pearly peach dun." He has four white socks over pink hooves, and shading on his knees and hocks that matches the mane and tail and is a warm, light brown, a little darker than the body. She said he has eye whites, "plus a dot for a highlight on his iris." (In comparison, the "Dapper" gift run has no eye whites, only one sock, and three brown hooves. Except for the difference in the number of socks, "Dapper" and his test have the same pinto markings.) *Photos E & F, courtesy of Deana Sprague.*

## G

- Reddish-Gold Buckskin Polo Pony with white bandages, Test Model 883-08

Owner Marla Phillips said she bought this model on eBay in 2004 from the original owner, "Denise D.," who got it in 2001 while she worked for Hartland! It has rich, reddish gold buckskin color and black points. The model is not dappled, and its body color appears deeper than the gift-run "Cinnamon Dapple Bay" models. Marla reported the same belly number shown in the photo on eBay: 883-08. *Photo G, courtesy of Marla Phillips.*

**G**
Reddish-Gold Buckskin Polo Pony with white bandages, Test Model 883-08.

## H

- Buckskin (Dappled) Polo Pony with Teal-Green Wraps, Test Model

With its teal-green wraps, black points, and dappled, golden buckskin body color, this 9" series model is not similar to any production runs. The owner, Victoria Zanutto, said she bought it in around 2004 or a few years later than that. On its bottom is written: Test 1 of 1. *Photo H, courtesy of Victoria Zanutto.*

## I & J

- Black Pinto 9" Polo Pony, Test Model 883-09

This model was freehand-airbrushed in a black pinto that's about 65% white. It is mostly white over the neck and body, with four white socks and pink feet. It is black over the head, and on the chest, rump, tail, and upper legs. The bandages were not painted to contrast with the body. Collector Marla Phillips said she purchased it from Hartland designer Sheryl Leisure at a model horse show, possibly in 2005. *Photos I & J, courtesy of Marla Phillips.*

## K

- Dapple Grey (Pearled) 7" Tennessee Walker Stallion, Test Model 684S-04

In the December 2005 test color sale, this stallion with pearled finish, dark mane and tail, four white stockings, and a masked star was $45. The company called him: pearl dapple grey. The production run dapple grey (684S-03, September 2002) was not pearled. *Photo K, courtesy of Eleanor Harvey.*

**H**
Buckskin (Dappled) Polo Pony with Teal-Green Wraps, Test Model.

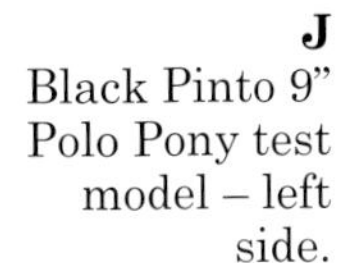

**J**
Black Pinto 9" Polo Pony test model – left side.

**I**
Black Pinto 9" Polo Pony, Test Model 883-09.

**K**
Dapple Grey (Pearled) 7" Tennessee Walker Stallion, Test Model 684S-04.

## L

- Pale Dun Polo Pony, Test Model 883-03

This light, reddish dun with dorsal stripe and leg bars was $75 in the December 2005 test sale. *Photo L, courtesy of Eleanor Harvey.*

**L**
Pale Dun Polo Pony, Test Model 883-03.

**M**
The Number 7 Palomino (Pearled) Polo Pony, Test Model 883-07, has golden pasterns.

**N**
The Number 12 Palomino (Pearled) 9" Polo Pony, Test Model 883-12, has white pasterns below the white leg wraps.

**Almost Twins.** The next two models, palomino Polo Ponies, look nearly alike, but one has white pasterns (low socks) and the other does not. To compare their golden color, they'd need to be seen side-by-side in person or in the same photo.

## M & O

- Palomino (Pearled) Polo Pony w/ White Pearl Wraps – Test Model 883-07

This test Polo Pony is pearled palomino, with pearl white leg and tail wraps. Owner Victoria Zanutto called the wraps, "white pearl." The hooves are pink. The pasterns are golden. (This model has no socks.) *Photos M & O, courtesy of Victoria Zanutto.*

**O**
Number 7 Palomino (Pearled) Polo Pony, Test Model 883-07 – *left side.*

## N & P

- Palomino (Pearled) 9" Polo Pony, Test Model 883-12

"She is a soft, golden, slightly metallic palomino," owner Melanie Teller said. The wraps, pasterns, fetlocks, mane, and tail are pearl white; the hooves are peach; the muzzle is black. She's "pearled all over except for the insides of her ears, her muzzle and nostrils, and eyes," she said. This model was the trophy for the Overall Hartland Champion at Kitty's No Snow Model Horse Show, January 11, 2003, in Arcadia, California. *Photos N & P, courtesy of Melanie Teller.*

**P**
Number 12 Palomino (Pearled) 9" Polo Pony, Test Model 883-12 – *left side.*

**Production Palomino was Pale.** Hartland did produce a regular-run palomino Polo Pony with white wraps, 883-05 in September 2002, but it is not pearled at all, and is paler than these two rich and gleaming test models.

## Q

- Cream (Dappled and Pearled) 7" Tenn. Walker Mare, Test Model 684-13

In the December 2005 test sale, this mare with blue eyes, pink muzzle, and pink feet (hooves) was called: dapple pearl cremello. The price was $25. The knees and hocks are a little darker shade of cream than the body. In addition to the belly number, there is writing on the bottom of the right front hoof. It is hard to read, but looks like a "C" and "Sond."

**Q**
Cream (Dappled and Pearled) 7" Tennessee Walker Mare, Test Model 684-13.

## R

- Dapple Grey (Pearled) 9" Polo Pony with white wraps, Test Model

Owner Eleanor Harvey said this was a test from 2001; probably, it was the first test model Hartland sold on eBay. That was a pearled dapple grey auctioned on May 28, 2001 (winning bid unknown). This test has dark hooves and points, except for white pasterns. In contrast, the production dapple grey with white wraps (March 2002) has pink feet, blacker points, and is not pearled. The test's partial number is: 883-. *Photo R, courtesy of Eleanor Harvey.*

**R**
Dapple Grey (Pearled) 9" Polo Pony with white wraps, Test Model.

## U & V

- Golden Chestnut Pinto (Pearled) 7" TWH Mare, Test Model 684M-12

This mare has a sheen that is not quite metallic, but might be called pearled, according to her owner, Melanie Teller. She has pink hooves, a flaxen mane and tail, a golden chestnut base color, and overo pinto markings: a large, marking on each side, smaller spots on the neck, and a bald face.

In slightly imperfect condition, she was offered in the December 2005 test sale for $25, but apparently did not sell, and was donated as the trophy for the Hartland Division Reserve Champion at the Jamboree Winter Challenge model horse show, Feb. 28-March 1, 2009, in Mariposa, California. *Photos U & V, courtesy of Melanie Teller.*

## W

- Dark Gray Appaloosa 9" Polo Pony w/ Red Wraps, Test Model

This appaloosa has only a few black spots on the blanket area. Owner Victoria Zanutto called the red wraps on the dark-gray-to-black body, "really dramatic," adding, "He's one of my favorites." The number written on the model's underside is faded, but appears to be: 883-01-T. *Photo W, courtesy of Victoria Zanutto.*

**S**
**Heads Up –** On the 9" series Five-Gaiter, a vein is prominent between the eye and the nostril. This one, the Silver-Gray 9" Five-Gaiter Test Model 881-10, with a star on its face, is also in Chapter 14 (photo I). *Courtesy of Victoria Zanutto.*

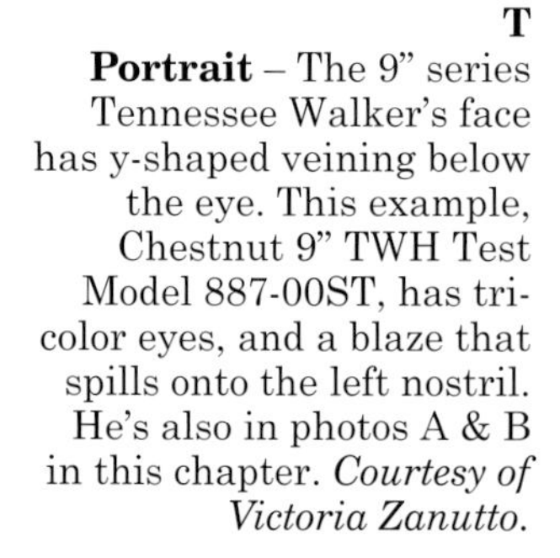

**T**
**Portrait** – The 9" series Tennessee Walker's face has y-shaped veining below the eye. This example, Chestnut 9" TWH Test Model 887-00ST, has tri-color eyes, and a blaze that spills onto the left nostril. He's also in photos A & B in this chapter. *Courtesy of Victoria Zanutto.*

**U**
Golden Chestnut Pinto (Pearled) 7" TWH Mare, Test Model 684M-12.

# TEST MODELS NOT SHOWN: 9" TENNESSEE WALKERS AND 5" AND 7" FAMILY SERIES HORSES

## 9" SERIES TENNESSEE WALKERS – TESTS:

- Pale Palomino (Pearled) 9" Tennessee Walker – Test 886-

A test-color 9" series Tennessee Walker in "pale golden pearl palomino" with four white socks and pink hooves was auctioned on eBay on May 11, 2004. With two hours to go, the high bid was $46.

- Clear Plastic (Unpainted) 9" Tenn. Walker – Test 886-

**V**
Golden Chestnut Pinto (Pearled) 7" TWH Mare test model – *left side.*

**W**
Dark Gray Appaloosa 9" Polo Pony w/ Red Wraps, Test Model.

Hartland's auction on eBay described this model as clear plastic with a slightly green tinge. It explained that the plastic is clear until white dye is added, and that the first molded pieces sometimes come out without the dye. This model was an instance of that. The bidding was at $31 with two hours to go on May 11, 2004.

## 7" SERIES TENNESSEE WALKER (TWH) STALLIONS – TESTS:

- Greenish-Black 7" TWH Stallion – Test 684S-007T

The May 2002 test-sale catalog said this model was meant to be blue-black, but came out with "deep sea green" highlights. It has four white socks, mostly low. The price was $150.

- Flea-bit Grey 7" TWH Stallion – Test 684S-

A test-color TWH stallion described as "Chestnut Flea-bit" was painted November 9, 2000, and sold on eBay on June 4, 2001. With four days to go, the bidding had reached $31. The color is white-grey with slight chestnut shadings and tiny dark flecks. The muzzle is much darker than the body. The hooves are pale.

## 7" SERIES TENNESSEE WALKER (TWH) MARES – TESTS:

- Flea-bit Grey 7" TWH Mare – Test 684M-

A test-color TWH mare was painted white-grey with chestnut shadings and dark specks on November 9, 2000, and sold on eBay June 4, 2001. The company called it "chestnut flea-bit," but the chestnut color is limited to shadings, which are especially noticeable on the knees and hocks. The muzzle and feet look black. The mare differs from the test-color Flea-bit Grey stallion, which has pale hooves. Like the stallion, it was bid up to $31 with four days to go.

- Golden Palomino 7" TWH Mare – Test 684M-06

This mare with pale hooves, freehand socks, and no dappling was $25 in the December 2005 test sale.

- Palomino (Dappled) 7" TWH Mare – Test 684M-08

This test with pale hooves, white socks, and dappling sold for $20 in the December 2005 test sale.

- Shaded Bay (Dappled) 7" TWH Mare – Test 684M-07

The legs are solid black with no white on this dappled bay test. The bay production run was not dappled. The test was $35 in the December 2005 test sale.

- Buckskin (Dappled) 7" TWH Mare – Test 684M-03

She has dapples, a dorsal stripe, and solid black legs with no white. The production run buckskin 7" TWH Family (684-03, February 2002) is not dappled. The mare was $25 in the December 2005 test sale. Her number is like the production run's, but with an "M" for "mare" added.

- Buckskin 7" TWH Mare – Test 684-17

This test has a dorsal stripe and leg barring, but no dapples. The buckskin production run family (684-03) does not have any striping. In the December 2005 test sale, this mare was $25.

- "Champagne" Overo Pinto TWH Mare – Test 684M-18

This test has a large, overo pinto marking on her right side (the left side was not pictured in the sale), one freehand sock, a bald face, pink nose, and three dark hooves. "Amber champagne" was the factory name, but the color is either gold champagne or light chestnut with lighter mane and tail. In the December 2005 test sale, she was $25.

## 7" Series Tennessee Walker Foal – Test:

- Chestnut Tobiano Pinto 7" Tenn. Walker Foal – Test 684F-

Hartland auctioned a chestnut pinto TWH Foal on eBay on May 30, 2001. The bidding had reached $76 with 31 minutes to go. This test model was a tobiano pinto with white over the tail, the back, and most of the neck. The brown paint was freehand airbrushed; the pinto pattern was not masked. The model was painted November 9, 2000.

## 5" and 7" Family Sets – Tests

**Family Series Horses Pictured in Black-'n'-White.** The following test color models were shown in black-and-white photos in two issues of the company newsletter, Winter 2005 and Spring 2005:

- 7" Three-Gaited Saddlebred Mare and Foal in bay tobiano pinto; the mare has black on her mane and tail, but the foal appears to have brown on its mane and tail. (I think this set is different from the one displayed at Jamboree in 2009.)
- 7" Arabian Family (3 pc) in bay tobiano pinto.
- 7" Morgan Family (3 pc) in dappled bay.
- 7" Tennessee Walker Family (3 pc); the stallion might be chestnut roan, but the mare, with black mane and tail, is bay roan. The foal looks roan, too, but shows less contrast between head and body color. Production-run families were issued in bay roan and chestnut roan, in 2006, so this test family "went somewhere."
- 5" series Quarter Horse mare and foal set in dark (probably black) appaloosa. The mare has a white blanket with dark spots in it. The foal has a flurry of white spots on its rump.
- 5" series Thoroughbred mare and foal in buckskin. The mare has four socks; the foal has hind socks.
- 5" series Arabian mare and foal set in "rose grey." They are near white with slight brownish shadings and black mane and tail.

**Family Series Test Horses Displayed at Show in 2009.** The following 7" series and 5" series test models, seen in color photos by collector Melanie Teller, were displayed at the Jamboree Winter

Challenge Model Horse Show at the Mariposa County Fairgrounds in Mariposa, California, February 28-March 1, 2009. The models were in the possession of Sheryl Leisure, horse director for Hartland 2000.

## 7" Family Series Horses – More Tests:

- Arabian Family in chestnut with golden chestnut mare, pale golden chestnut foal, and chestnut tostado stallion. (Chestnut tostado means the mane and tail are a darker brown than the body.)
- Arabian Family (3 pc) – bay like the Tennessee Walker Family that was released. The foal's color is paler than the adults.
- Arabian Family (3 pc) – bay tobiano pinto
- Arabian mare and stallion – bay going grey. (No foal?)
- Arabian mare and stallion – grey with black points. (No foal?)
- Arabian stallion (alone) – golden chestnut
- Arabian mare (alone) – light grey with dark hooves
- Morgan Family (3 pc) – pale palomino
- Morgan family (3 pc) – bay
- Morgan family (3 pc) – liver chestnut
- Saddlebred mare and foal – light chestnut pinto, the color of "Copper King"
- Saddlebred mare and foal – bay
- Saddlebred mare and foal – black pinto (70% white)

## 5" Series Mare and Foal Sets – More Tests:

- Arabians – black with white socks
- Arabians – grey mare and bay-shaded grey foal
- Arabians – white-grey mare and chestnut foal
- Quarter Horses – bay
- Quarter Horses – black appaloosa
- Thoroughbreds – bay
- Thoroughbreds – white mare and gray-grey foal

# CHAPTER 17
# TESTS
## SUNLIT SADDLEBREDS

Among the unique ("test color") models were many Saddlebreds, including palominos, the "sunlit Saddlebreds." This chapter also describes test Tinymites, 9" series horses with reins – some with Indian riders, and illustrates more 11" Arabians.

**A**
**Unforgettable Eyes** – The Bay Roan 11" Arabian, Test Model 901-015T, has tri-color eyes with a brown iris, and a knowing look.

**B**
The Bay Roan 11" Arabian, Test Model 901-015T, is handsome from head to foot.

**C**
Shaded Bay Pinto 11" Arabian, Test Model #901-T 017.

**D**
Shaded Bay Pinto 11" Arabian test model – *left side.*

**E**
The horse from the Dapple Grey 9" Semi-Rearing Horse and Indian Test Set 813-008 is the mane-down version of the Semi-Rearing horse.

**F**
The Dapple Grey 9" Semi-Rearing Horse and Indian Test Set 813-008 includes a removable saddle and elaborately painted headdress and shield.

## A & B

- Bay Roan 11" Arabian, Test Model 901-015T

Hartland Horses donated this model as a reserve champion trophy at the Jamboree Summer Challenge model horse show at the Jamboree in Pomona, California, October 9-10, 2004. He has a whitish body with brown shadings, and except for a white face marking, has a solidly dark brown head, which is typical for bay and chestnut roans. He has black points except for four white socks. He has peach hooves, and a star and snip on his face. Owner Nerissa Pospychala said he has tri-color eyes with a white front corner, brown iris, black pupil, and white back corner. *Photos A & B, courtesy of Nerissa Pospychala.*

## C & D

- Shaded Bay Pinto 11" Arabian, Test Model #901-T 017

This model's body shades from medium bay to dark bay. He has high white stockings, pink feet, no face markings, and tri-color eyes, according to owner Nerissa Pospychala. His tobiano pinto pattern was used for a black pinto gift run and for production runs in black, chestnut, and bay pinto, but the production bay pinto was a lighter brown. He was a show prize awarded at the Jamboree Summer Challenge model horse show at Jamboree in October 2004. *Photos C & D, courtesy of Nerissa Pospychala.*

## E & F

- Dapple Grey 9" Semi-Rearing Horse and Indian – Test Set 813-008

Hartland Horses auctioned this unique, test model Indian horse-and-rider set on eBay on February 17, 2007. The winning bid was $359.99.

The horse is Semi-Rearing with Mane Down, in dapple grey with black mane and black tail, except for a white tip. The knees, hocks, hooves, and bridle were painted black. The saddle and plastic reins are black, too. The rider is the Chief Thunderbird shape with brownish skin, pale tan leggings with brown fringe, and accessories in black, white, and silver color. The set includes a shield, knife, spear, rifle, bow, and tomahawk. *Photos E & F, courtesy of Judy O'Bannon.*

## G

- Pale Palomino (Dappled and Pearled) 9" Five-Gaiter, Test Model 881-017T

In the May 2002 test sale, this horse was $175 and described as: light gold pearl palomino with some very light dapples, four socks, bald face, and pinkish-gold hooves. The white areas were obviously pearled and the whole model appears to glow. This color should have been a production run. *Photo G, courtesy of Pamela Pramuka.*

**G**
Pale Palomino (Dappled and Pearled) 9" Five-Gaiter, Test Model 881-017T.

## H

- Deep Palomino 9" Five-Gaiter, Test Model 881-04

Owner Anni Koziol said this horse is, "matte palomino, very rich in color." In the 2005 test sale, it was $65 and called: taffy shaded palomino with freehand socks and a freehand star. The feet are peach. The eyes are black with a white back corner and highlight. *Photo H, courtesy of Anni Koziol.*

**H**
Deep Palomino 9" Five-Gaiter, Test Model 881-04.

**Greys with snowy manes.** The next four test Saddlebreds are grey with a white mane and a white, or mostly white, tail.

## K

- Shaded Grey (Pearled) 9" Five-Gaiter, Test Model T881-023

This grey with dark gray or black shadings has a matte finish, white mane and tail, black hooves, and black on the knees, hocks, and muzzle. The horse has four white socks, but dark shading noticeably spills down the front of the cannon bone on the right foreleg. There are no tiny specks of color in the coat. The eyes are solid, dark gray according to the owner, Anni Koziol. In the December 2005 test sale, this model was called: pearl gray with white mane and tail. The price was $75. *Photo K, courtesy of Anni Koziol.*

**Greys with midnight manes.** The next two test Saddlebreds are grey with largely black points.

## I

- Number 21 Rose Grey (Dappled) 9" Five-Gaiter, Test Model 801-021T

This is, "A very matte rose grey with dapples and dorsal stripe," owner Anni Koziol said. This model has an essentially grey body with faint brown shadings framing small dapples. The mane, lower legs, and hooves are black. The right side of the tail is entirely black, but the left side has a light tail tip. The December 2005 test-sale catalog said the model has dark shading, but no white socks, and that is correct. The price was $75. *Photo I, courtesy of Anni Koziol.*

**I**
Number 21 Rose Grey (Dappled) 9" Five-Gaiter, Test Model 801-021T.

## J

- No # Rose Grey (Pearled) 9" Five-Gaiter, Test Model

"Pearled rose grey with a slight gloss," was the description by owner Anni Koziol. The mane, tail, lower legs, and hooves are black, except the tip of the tail is light. The model has black eyes. It is pearled, but not dappled whereas the production-run rose grey models (881-03), are dappled, but not pearled. This test is not numbered on the bottom. *Photo J, courtesy of Anni Koziol.*

**J**
No # Rose Grey (Pearled) 9" Five-Gaiter, Test Model.

## L

- Flecked ("Flea-bit"), Shaded Grey 9" Five-Gaiter, Test Model 881-02

This grey has both gray shadings and tiny dark specks in its coat; its mane and tail are white. The knees, hocks, hooves, and muzzle are black. The eyes are black with eye whites and a white dot highlight. Owner Anni Koziol said that the writing on the bottom of the right rear hoof looks like: 1/9/00, a date that precedes the formation of Hartland 2000 by several months. In January 2012, Sheryl Leisure said that the earliest she painted any test colors for Hartland 2000 was fall of 2000. The date was probably 11/9/00, but marker pen fades. *Photo L, courtesy of Anni Koziol.*

**K**
Shaded Grey (Pearled) 9" Five-Gaiter, Test Model T881-023.

**L**
Flecked ("Flea-bit"), Shaded Grey 9" Five-Gaiter, Test Model 881-02.

## M

- Dapple-Grey 9" Five-Gaiter with Dark Tail Tip, Test Model (No #)

Of the grey Saddlebreds with a white mane, this is the only one with dapples, and the only one with pink, rather than black, hooves. Owner Anni Stapley-Koziol said that the body is very matte, and the mane and tail look pearled. The tip of the tail is black. The face has a white blaze and pink snip. Written on the belly, in worn letters, is: Test. *Photo M, courtesy of Anni Koziol.*

**M**
Dapple-Grey 9" Five-Gaiter with Dark Tail Tip, Test Model (No #).

## N

- Flea-bit Grey (Mainly White) 9" Five-Gaiter, Test Model T881-022

The owner described this Saddlebred as "white flea-bit grey with some slight chestnut shadings" and black hooves. It has tri-color eyes: brown and black with eye whites in the front and rear corners. On the 2005 test-sale list, the color was called: flea-bitten chestnut, black hooves. The price was $75. *Photo N, courtesy of Anni Koziol.*

**N**
Flea-bit Grey (Mainly White) 9" Five-Gaiter, Test Model T881-022.

## O

- Chestnut (Dappled) 9" Five-Gaiter, Test Model 881-03

"Matte, dappled chestnut that looks almost copper orange," is how the owner, Anni Koziol, described this model. It has a darker brown mane and tail, pink hooves, mostly white face-front, and eyes that are black with a white back corner and highlight. In the December 2005 test sale, it was $65 and called: dapple chestnut with four freehand socks. *Photo O, courtesy of Anni Koziol.*

**O**
Chestnut (Dappled) 9" Five-Gaiter, Test Model 881-03.

## TEST MODELS NOT SHOWN: 9" FIVE-GAITERS, 9" RIDER SERIES, AND TINYMITES

## 9" SERIES

## FIVE-GAITED AMERICAN SADDLEBREDS - TESTS:

- Black with White Points 9" Five-Gaiter – Test T881-024

This test apparently inspired the October 2001 production run of the same color, but the production run, 881-02, is pearled, and the test is not. The test has a black body, white points, and peach hooves. Its eyes are brown with eye whites in the front corners, according to its owner, Anni Koziol. Written on the bottom of the right hind foot is: Charcoal; written on the right front hoof bottom is: T881-024.

- Palomino 9" Five-Gaiter with masked socks – Test 881-01

This model dated "11/9/00" was $65 in the December 2005 test sale. (It differs from the palominos shown in this book.) This test was not pearled, but the production palomino (March 2001) was pearled. The test and the production model both have the same number: 881-01.

**Test 9" series Saddlebreds in Newsletter.** Two more test 9" Five-Gaited Saddlebreds appeared in small, black-and-white photos in the company newsletter with no color description. Neither of them matches the appearance of other known models.

- A Five-Gaiter in a medium-light body color with dark mane and dark upper half of the tail, four stockings, and dark feet was sold by Hartland on eBay in about the first quarter of 2003 (Winter 2003 issue, which arrived in April 2003).

- A Five-Gaiter in a light color (palomino?) with very black hooves, four white stockings, light mane and tail, a little shading on the knees, and a dark muzzle was proposed to be sold on eBay by Hartland around mid-2003 or later (Spring 2003 issue, which arrived in July 2003).

## 9" RIDER SERIES - TESTS

**Test 9" Horse & Indian Rider Sets.** Hartland 2000 numbered its Indian horse-and-rider sets with the prefix "813," the number used for the Chief Thunderbird sets in the 1950s and 1960s.

- Grey (?) Pinto Semi-Rearing Horse & Indian – Test Set (Auctioned) 813-

This horse's body is a light color, such as gray or buckskin, and its pinto pattern resembles the Longley and Cochise pintos from the 1950s, 1960s, and 1980s. It has a dark mane, dark shading on the knees, and medium-to-dark hooves. The bottom two-thirds of the tail looks white while the upper one-third is dark. That's all I can tell from a one-inch high, black-and-white photo in the 2001 Jamboree program. The mold is Semi-Rearing with Mane Up (and probably Plain Tail). This Indian rider set was Lot #20 in the open auction at Jamboree; the minimum bid was $75.

- Buckskin Pinto Horse & Indian – Test Set (Raffled) 813-

A test-color Hartland Indian rider set with a buckskin pinto horse was a Friday raffle item at Jamboree in June 2001. The Jamboree program did not picture the set or identify the horse's mold, but it was probably semi-rearing, and may have been similar to the set (above) that was auctioned at the same event.

- Bay Pinto Horse & Indian – Test Set 813-

A test-color Hartland Indian rider set with a bay pinto was raffled on Saturday during Jamboree in June 2001. The set was not pictured and the horse's shape was not identified in the Jamboree program, but it was probably semi-rearing.

- Black (Pearled) 9" Semi-Rearing Horse & Indian – Test Set 813-604

Hartland sold a test, Semi-Rearing 9" horse (with Mane Up and Plain Tail) in pearled black with white stockings, bald face, and bridle painted red, on eBay on October 25, 2001. It was part of a set with an Indian rider with loincloth painted red and a group of plastic, Indian accessories. The number written on the bottom of the Indian was: 813-604.

- White (?) Horse & Indian – Test Set 813-

The company's Winter 2003 newsletter included a small, black-and-white photo of an Indian rider set with a Semi-Rearing horse with Mane Up and Plain Tail. It said the set was among test models recently sold on eBay (prior to the newsletter's arrival in April 2003). The horse looks white with black feet and relatively dark war paint. The accessories look more elaborately painted than on the 1980s Steven Indian set, which also used a white horse of the same mold.

- Palomino Pinto 9" Semi-Rearing Horse & Indian –Test Set 813-007T

This Semi-Rearing horse (with Mane Up and Plain Tail) is mainly golden-yellow with smallish, white pinto spots and war paint in shapes and locations like those seen on early 1960s Semi-Rearing Chief Thunderbird sets. The white, pinto spots are on the left hip, neck, and shoulder and right neck and hip. The war paint includes a turquoise sun symbol on the horse's left side; and a red snake, yellow dashes, and black deer prints on the right side.

Paired with Chief Thunderbird in a turquoise loincloth, the set was offered at silent auction in the May 2002 test sale, but apparently not actually sold until May 11, 2004, on eBay (or else there were two, identical sets).

The horse was equipped with a brown plastic rein, and blanket saddle painted black; its bridle was painted brown. The set included accessories – headdress, spear, knife, tomahawk, rifle, shield, and bow – painted mainly in black-and-white with some brown. The belly number in the catalog was: 803-007T, but was, probably, a typographical error, and meant to be: 813-007T.

**Test 9" Rider Series Horses on a Color Sheet.** Test-color, 9" rider-series horses were displayed in color in a mailing dated May 6, 2004. Each of the six horses was pictured with a black, plastic-lace rein and a black cowboy saddle (presumably, with rifle hole; only the left side was shown). The molded-on bridle headstalls were painted black, except on the black horse, whose headstall was painted brown. Each horse was molded in white plastic, and white areas appeared to be the unpainted plastic, not white paint that was added. Three were the Chubby Standing/Walking mold, and three were Semi-Rearing with Mane Down (abbreviated as: MD). The six are:

1. Chubby – gray appaloosa with black points and white blanket that extends forward almost to the withers. Six black spots are visible in the white blanket, on the left side of the haunches.
2. Chubby – reddish-brown chestnut with four white socks.
3. Chubby – coffee dun with black points. A model like this was issued as the production run named "Java Joe" in April 2005. (Note: The three test Chubbies above all have their correct tail. At least some of the "Java Joe" production models were issued with the wrong tail.)

4. Semi-Rearing (MD) – gray appaloosa with black points and a white blanket that extends forward to almost the withers. Six black spots are visible in the white blanket, on the left side of the haunches.
5. Semi-Rearing (MD) – black with no white. The only thing that isn't black is the bridle headstall, which is painted brown.
6. Semi-Rearing (MD) – solid bay. This model is chocolate brown with black points and some black shading on the muzzle. (No Semi-Rearing Horses with Mane Down have been issued as production runs, but the solid chocolate-bay color was used for "Cocoa," a Mane-Up Semi-Rearing Horse production run in December 2007.)

**Rider Series Horses Pictured in Black-'n'-White.** The Winter 2004 company newsletter, which arrived in May 2004 along with the above color sheet, included black-and-white photos of five more test colors on 9" rider-series horses, each wearing a dark-colored cowboy saddle. Each was a Semi-Rearing Horse, but some with Mane Down (MD) and others with Mane Up and Plain Tail (MUPT). They are:

1. Semi-Rearing (MD) – white with black points. (I informed Hartland that this model, although rare, had already been made: in the 1960s. I acquired one after *Hartland Horsemen* [1999] went to press.)
2. Semi-Rearing (MD) – gray appaloosa in the same pattern described in two models above.
3. Semi-Rearing (MUPT) in the same gray appaloosa color as above.
4. Semi-Rearing (MUPT) – bay with four white stockings; the mane, tail, knees, hocks, and hooves are black.
5. Semi-Rearing (MUPT) – coffee dun. This test model led to a production run: "Cuppa Joe," the 2005 Club Model.

## Tinymites: Six Breeds 3" Long - Tests:

- Tinymite 5 pc Set – Test Set (Auctioned)

A test set of five Tinymites was Lot #20 in the Jamboree Benefit Auction at Jamboree in October 2004. Each of five breeds – there is no draft horse – is a different color, and different from the production colors. The Jamboree program showed only a black-and-white picture with each horse one-half inch high. The Arabian is pale (probably palomino); the Tennessee Walker has an all-dark head so would be roan or, as the company often painted these colors, dun or grulla. Numbering was not mentioned, but if I assigned a code, it would be TM-53T.

- Tinymite 5 pc Set – Test Set (Raffled)

A different, test-color set of five breeds of Tinymites was part of the 2004 Jamboree Raffle (as opposed to the Jamboree Auction). The Arabian has dark points except for four white socks, and has an all-dark head, so is probably a roan (or a dun or grulla as this company would often paint those colors). The Jamboree program included only a small, black-and-white photo. Each breed is a different color, and different from the auctioned test set and the production colors.

There was a raffle each day for three days, and the program said it would include "Test sets [plural] of Hartland Tinymite horses." That makes it sound like more than one Tinymite set was raffled. If so, what the other one(s) looked like, if not like the picture, I do not know. My assigned code for the test set that is pictured in the program is: TM-54T.

## Chapter 18

# Other Unusual Models

This chapter corrals models that were not unique, but were unusual, and were not gift editions, but were sold. Values are included. Some were from 2000-2007, and some were from the previous Hartland company, Steven Manufacturing (1980s-1990s). Unpainted, white models used in painting workshops and a painting contest are also mentioned.

**A**
The Gray Swirl 9" Tennessee Walkers, 887-GS, are unpainted. The color is molded in.

**B**
Each Gray Swirl 9" Tennessee Walker was different. This one is mainly lighter shades of gray.

**C**
The tail is entirely dark on the right side of the Gray Swirl 9" Tennessee Walker in photo B.

**D**
Some 1990s Semi-Rearing horses with mane down, molded in yellow plastic, were sold unpainted.

## A, B, & C

- Gray Swirl (Unpainted) 9" Tennessee Walker, 887-GS; August 2002; Quantity: probably fewer than 10

No two of the Gray Swirl Tennessee Walkers look exactly alike because they were molded in plastic that is a random mix of white and shades of gray, and maybe even black. No paint was applied to them. They are glossy because that's how the plastic comes out of the mold: smooth and shiny. They have smoothed seams. (The plastic is cellulose acetate, the better plastic. It was not styrene.)

The Gray Swirl was a run of unknown size; probably, fewer than 10. I know of two collectors besides myself who have one, and two more Gray Swirls I saw on the Internet could be in addition to our three. One had a black or very dark gray tail and the dark color almost all the way up his left hind leg.

Collector Eleanor Harvey said she bought one from a "grab bag box jumbled with odds and ends" at Jamboree in August 2002. She said the teenagers manning the sale were unable to provide any information except the price. My guess would be that they were molded by Hartland 2000, rather than by the previous Hartland company.

More than a year later, in December 2003, I bought a Gray Swirl from the company website. A customer service person, Lynn, said it had been shown at the website by mistake. She wrote, "We don't ever plan to sell any of them from the website as they are all different, and it would be too hard to pick out specific patterns for people. We had only a few, and sold most of them a few years back at the Jamboree."

When I said I understood that they were all different, she agreed to sell me one, for $25 plus shipping. On the receipt, the company called the color: grey swirl. Collector Eleanor Harvey calls it: marbled gray. *Photo A, courtesy of B & K.*

Value in "as is from factory" condition: $35.

## D

- Solid Yellow (Unpainted "Trigger") Semi-Rearing Horse (with mane down); Molded in the 1990s; Quantity: About 10 or more

On May 11, 2004, Hartland Horses sold a solid yellow Semi-Rearing horse on eBay, calling it "very unique" and a "test model." It actually had no paint on it, and the auction said that it had been molded in the 1990s by Steven Manufacturing, the previous Hartland company.

The auction said, "When molding, the plastic comes clear, and so Steven added yellow dye so that they had a base coat of golden yellow to use for Trigger models back in 1993-1994." (Note: The full time span in which Steven issued Roy Rogers and Trigger was 1992-1994 [*Hartland Horsemen*, p. 68].)

The description also said that the model had rough seams because it was "not trimmed, but simply assembled straight from the molding machine." With two hours to go, the bidding had reached $169.38. As it turns out, there was more than one of these, but collectors had to wait several years.

When I bought one on November 14, 2010, seven or eight of the Unpainted Yellow Semi-Rearing horse were available on eBay for $35. (They were called: Hartland Steven unfinished Roy Rogers horse Trigger.) On November 16, 2010, five remained. The auction by Horse-Power Graphics noted that, "Starting with tinted plastic meant fewer coats of paint" were needed. In June 2011, three were left, and in January 2012, two remained. (The plastic is quality plastic, not styrene.)

Rider-series horse mold numbers were described in *Hartland Courier*, Spring 2001. This horse's shape, Semi-Rearing with Mane Down, is mold #870.

Value in "as is from factory" condition: $35.

## E, F, & G

- Styrene "Buttermilk" (9" rider series Chubby horse); Quantity: 8 or 9

This is Dale Evans' horse, "Buttermilk," in styrene plastic. The production run released in 2005 was cellulose acetate, the better plastic. However, according to Horse-Power Graphics, the manufacturer molded and painted "a small batch (8 or 9)" of them in styrene plastic. In December 2010, HPGraphics was selling them on eBay for $34.99 plus $9.95 shipping. Three were left on May 20, 2011.

The paint used on the styrene and cellulose acetate ("Tenite") "Buttermilks" resulted in both of them photographing as the same color, but in person, the styrene model is glossier and its color is slightly rosier. Both versions of "Buttermilk" are pale buckskin or creamy dun. *Photo E, courtesy of Melanie Teller.*

Value, as from the factory (NM to Mint): $35.

- Clear Plastic Roy Rogers and Dale Evans Unassembled Sets; Quantity: 6 or slightly more

**Not Shown.** On December 31, 2010, Horse-Power Graphics was selling clear plastic test shots of Roy Rogers, Dale Evans, Trigger, Buttermilk, and Bullet figurines on eBay. The group of unassembled, untrimmed, and unpainted pieces included front and back halves of Roy and Dale, and left and right sides of the three animals plus Buttermilk's tail, two Roy Rogers saddles, and the hats and guns for Roy and Dale. These parts are quality plastic, not styrene.

The auction stated that several samples in untinted (transparent) plastic were made for inspection preceding the production run in normal plastic. The production models were released in 2005, but the clear samples probably date to 2003, the date on the stickers on the production sets, whose release was involuntarily delayed.

The auction said there were two Roy Rogers saddles because "the molder did not make any clear Dale saddles." (For the 2005 release, Dale had a cowboy saddle with rifle hole. That saddle was not on the same mold with Roy and Dale rider parts. However, two Roy Rogers saddles were on one of the applicable molds.)

Six of these groups of Roy and Dale parts were available on eBay on December 1, 2010. The asking price per lot was $250 plus $9.57 shipping. Six months later, on May 20, 2011, five were, reportedly, still available.

Before being grouped, there were three separate auctions of these clear parts in November 2010. Seven bidders fought over "Roy Rogers and Trigger," which sold for $128.50. "Dale Evans and Buttermilk" also sold for $128.50. "Bullet," the dog by himself, auctioned for $99 with 10 bids placed by six different bidders. The opening bid for each auction was $9.99.

**E**
In 2003, some sample "Buttermilk" Chubby horses were molded in styrene plastic (the cheaper plastic). The production run used better plastic.

**F**
The sample "Buttermilk" in styrene *(right)* and the production "Buttermilk" in the better plastic *(left)* were painted alike.

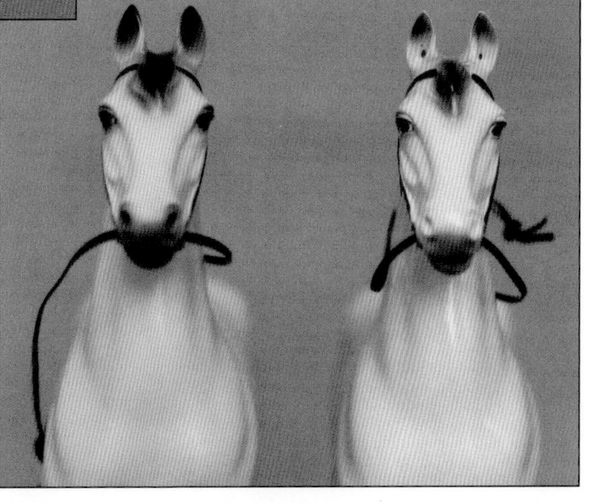

**G**
The styrene "Buttermilk" *(right)* has a pin-sized vent hole molded in each ear. The production run of "Buttermilk" *(left)* has normal ears.

Then, the group was split up again. The two riders with their horses and accessories were $85 each in November 2011. As collector Sharon Elosser pointed out, Bullet, the dog, was being sold separately, for $8. In January 2012, six each of the clear, unassembled Roy Rogers and Dale Evans sets remained for sale along with more than 10 of the German Shepherds. Sharon said she glued her Bullet together, but does not plan to paint it.

## Old Paint Sample Models from Steven Manufacturing

In November 2010, Horse-Power Graphics ("HPG") sold about 30 paint sample models of breed series horses and a few riders and rider-series horses. The samples were used as painting guides by Steven Manufacturing, the previous Hartland company, in the late 1980s and early 1990s. The opening bid for each horse was $4.99. Most of them auctioned for about $10-$20 each; the range was $5 to $39.

While they are "special" since they were the paint-color samples, entire mass-produced runs of models resemble them (in most cases). Their condition was described as "fair to rough." They got knocked around while they were in the Steven factory.

According to the auctions, the samples came with the purchase of Hartland from Steven in 2000, and were sent to "the contracted painting prototype studio in California" (in other words, HPG) to possibly use for inspiration (or just because horses belong where horses are being painted; the sports half of Hartland 2000 wouldn't have had much use for them). Later, when Hartland 2000 owed "the studio" money, the Steven paint samples became the property of HPG.

Since they are from another Hartland era, the Steven paint samples will be detailed in another book, but two examples are shown here.

### H

- Rusty Grey 9" Grazing Arabian Mare, Paint Sample

This model was the example followed by Steven Manufacturing for its #221 "Chestnut Flea Bit" Arabian Mare, "Fair Maiden," 1993-1994. I paid $9.49 for it in a November 2010 eBay auction. Sheryl Leisure painted some sample Hartland models for Steven Manufacturing in the early 1990s, and believes that she painted this one. In *Hartland Horses and Dogs* (2000), page 24, I dubbed the color, "Rusty Grey."

I think the 9" Arabian Mare mold is lovely, but grazing model horses in general are not popular. Steven had a lot of leftover stock of the "Fair Maiden" production model, new-in-box, and Hartland 2000 began selling them (through HPG) in December 2000. A decade later, HPG is still selling them. Nine of the mint, packaged Rusty Grey Grazing Arabian Mares were on eBay for $19.99 in June 2011, fully 17 years after their production ended! Wouldn't it be nice if opportunities like that were commonplace?

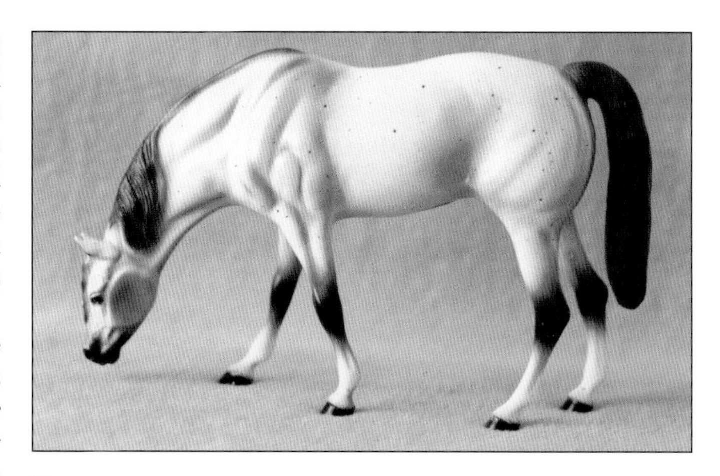

**H**
This paint sample model was used to guide the production of "Fair Maiden," the 1990s Steven #221 Arabian Mare in rusty grey.

### I

- Red Roan Appaloosa 9" Polo Pony, Paint Sample

In November 2010, Horse-Power Graphics' sale of Steven Hartland paint sample models included three different Red Roan Appaloosa Polo Ponies (Steven #228, 1988-1990. Paola Groeber's Hartland company, 1987-1990, produced this color, also. Paola, herself, painted some of the sample models for Steven in the late 1980s.)

It has the Steven factory tag still on it, as did some others in the sale. The tag reads, "#228 9" Polo Pony Leopard Appaloosa" although a leopard appaloosa would be light with dark spots all over.

With its deep, russet brown forehand and legs, this sample was the darkest of the three, and is cellulose acetate plastic (the better plastic, often referred to by a brand name, "Tenite"). The auction said it had "a few small rubs and nicks."

Lori Ogozalek's winning eBay bid was $13.00. The other two appaloosa samples did not have rubs. A lighter sample in "Tenite" sold for $24.49, and one in styrene was bid up to $22.07.

I think Lori's is the best looking of the three, but she paid the least. She said her winning bid was "really low because it was not an 'LSQ' model. For some reason, new collectors are stuck on '[it's either] perfect or it's a body,'" she said.

That "either/or" thinking omits any middle ground. "LSQ" means "live show quality; "PSQ" means "photo show quality." In a photo show, a model can get by with flaws that could hamper it at an in-person ("live") show. However, there is also "shelf quality," a model nice enough to admire at home. Most model horses are never entered in competitions. Some collectors intent on live showing, though, go overboard and consider even slightly flawed models as "bodies" useful only for repainting. Some of the models that get repainted would have pleased another owner, flaws and all. Perfection can be overrated. *Photo I, courtesy of Lori Ogozalek.*

**I**
Another paint sample model sold in 2010 was this one for the 1990s Steven #228 Red Roan Appaloosa Polo Pony.

## UNPAINTED, WHITE HORSES

Unpainted, white Hartland horses were not sold as a catalog item after 2000, but were released, so to speak, through painting workshops at Jamboree and a painting contest. Some of the participants were very talented. Hopefully, they signed their work so those models will not, later, be mistaken for a unique model from the factory!

**Workshops.** For the Air Brush Seminar by Caroline Boydston at Jamboree 2001, "All students got a little Hartland weanling foal to practice on" (according to Sheryl Leisure, in a message posted May 25, 2001, at Haynet, www.yahoo.com). Those would be the 9" series Weanling Foals; probably, they were old Steven stock from the 1990s.

The air-brush painting workshops at Jamboree in 2002, 2003, and 2004 also used, "a small Hartland model horse," according to the printed programs. If the Weanling was not used again, other "small" possibilities include the 7" series Tennessee Walkers and the Tinymites.

Each air brush painting workshop was $30. The class sizes were limited each year: to 20 in 2001, 25 in 2003, and 15 in 2004. In addition, the 2003 and 2004 Jamborees included a $10, children's model-horse painting class, limited to 25. Those classes used "a small-scale plastic horse model" of unspecified brand.

**Painting Contest.** The eighth annual "Tricked-Out Pony Contest" at the 2009 Winter Jamboree Challenge model horse show February 28-March 1, 2009, in Mariposa, California, used unpainted Hartland horses. Children painted a 7" Tennessee Walker stallion; adult contestants used an unpainted 9" series Polo Pony. The models had to be purchased in advance, and returned three weeks before the show. They could be reshaped so well as repainted.

Winners were announced at the show. The show and contest were sponsored by Horse-Power Graphics, Inc., with the models provided by Hartland Horses. After the show, the models were auctioned on eBay with 50% of the proceeds going to an unspecified animal charity, 25% to the sponsors, and 25% to the contestants.

I wonder whether any collectors bought the unpainted, white model just to keep it that way? It would look nice.

**Values.** Suggested values for unpainted white models: 9" Weanling Foal and 7" series: $7 if rough seams and $10 if smooth seams; 9" series: $10 if rough seams and $15 if smooth seams.

# HISTORY

**History of Hartland Horses Produced Since 2000.** The new Hartland company that formed in 2000 was unusual in that it was spread over more than one geographic location, and much of its work was contracted out. Hartland Collectibles, L.L.C., a Limited Liability Company registered in Missouri, was started by a group of investors for the purpose of manufacturing and selling sports statues and horse and western figurines. The company was headed by two managing partners: one with a business office in Shreveport, Louisiana, and the other, with a sales and marketing office in Hermann, Missouri. [Their names are withheld due to a request for anonymity. I am not obligated to omit their names, but have chosen to do so.] Their primary interest was the sports figurines, but from here on, this chapter will concern itself only with the horse and western line.

As explained in my previous Hartland books from Schiffer Publishing, there had already been a few Hartland companies with similar names. To prevent confusion, I will sometimes refer to the most recent company, whose models are the subject of this book, as: "Hartland Collectibles (2000)" or "Hartland 2000," rather than by the company's full and correct name, Hartland Collectibles, L.L.C.

In October 2000, the purchase of the horse and western line from Steven Toy of Hermann, Missouri, was complete. (Steven Toy was known as Steven Manufacturing Company while it made Hartland models in the 1980s and 1990s.) The purchase in 2000 included old Steven stock, both finished and unfinished. One of the first things Hartland 2000 did was to sell the old stock of new-in-box Steven horses from the mid-1980s through 1994. The sale was announced early in December 2000, and most of the models were sold out within a few days.

The sale included about six pieces each of 18 designs of 9" series Steven horses from the 1980s: nine designs in styrene plastic for $10, and nine designs in cellulose acetate plastic for $15. In addition, about 25 sets of Lady Jewel and Jade in bay (with rough seams) sold for $25; the same pair in dapple grey and black (with smooth seams) was $45. The price was $24 for the 7" series Tennessee Walker Mare and Foal set in palomino, and for the flea-bit, rusty grey 9" series grazing Arabian mare. (The 1980s-1990s models are pictured and described in *Hartland Horses and Dogs,* 2000.)

The sale was conducted from California, on consignment, by Horse-Power Graphics, Inc., a business owned by artists (illustrators) Sheryl Leisure and her husband, Brad. California was also where the new models from Hartland 2000 were painted (except for two rider sets manufactured in China). Hartland started a website in late November of 2000, and the sample models shown there were painted by Sheryl, a long-time model horse collector, customizer, sculptor, and promoter. From 1991-December 1999, she had published a monthly model horse "adzine" (consisting of copy-ready sales lists and other ads), and had done some model design and other art work for Breyer and for Peter Stone. She was still hosting her own annual, model horse convention, The West Coast Model Horse Collector's Jamboree (abbreviated as "Jamboree"). As an independent contractor, she began working for Hartland under her business name, Horse-Power Graphics, Inc., hereinafter often abbreviated as "HPG." Sheryl designed the horse colors for Hartland 2000, but usually did not paint the runs of models. A publicity photo showed two painters: young men, one of them wearing a cap backwards.

In December 2000, the Hartland company began taking orders; specifically, for its 2001 Horse and Western Club membership. Orders could be placed online or by mail. In December, the address was a post office box in Shreveport, Louisiana. The company name and address on the receipts enclosed with the models changed over time, so I'll put the exact name from the receipts in quotes in order to highlight the distinction. The first California address that "Hartland Collectibles" used from

January through March of 2001 was a post office box in Calimesa, California. Because of the three-state corporate location, the printable mail order form at the Hartland website in March 2001 included a line for residents of California, Louisiana, and Missouri to add state and local sales tax (8.25%, 7.5%, and 5.225%, respectively) to the purchase price of the models!

The receipts from April 2001 through October 2001 were from "Hartland Collectibles" at 700 E. Redlands Blvd., Ste. U-242, Redlands, California. "Denise D." [last name withheld at her request] worked for Hartland during the first year it was in business, at the office in Redlands. Her name appeared on emails notifying collectors of models for sale. A horse collector for many years – she had subscribed to my Hartland newsletter in 1994-1996 – she was able to bring something extra to her customer service role.

**First Model Arrives in 2001.** In *Horsing Around Magazine*, May/June 2001, I wrote, "The first models by the new Hartland company, Hartland Collectibles, L.L.C. (Limited Liability Company), are well painted and a wonderful awakening for a brand of model horses that had been dormant since November 1994."

The first horse from Hartland 2000 was a Polo Pony named "Chart the Course." It was the 2001 club model, and it arrived in February 2001. As I wrote in my own newsletter, *Hartland Model Equestrian News*, March-April 2001, this model's color was "both attractive and distinct from previous models." With the help of both of my books (*Hartland Horsemen*, 1999, and *Hartland Horses and Dogs,* 2000), the company was careful to not duplicate any of the exact model colors that had been done before. (Nearly one-fourth of the 1940s-1990s Hartland production models never appeared in company catalogs, but do appear in the books!)

The company was, at least initially, painting models that had been molded in the 1990s by Steven Manufacturing. About the models from February and March 2001, I wrote, "The new models were actually molded in 1994" (*Horsing Around Magazine,* May/June 2001). However, Hartland 2000 was also having new model bodies molded; reportedly, by a company in Kansas. Hartland's work was taking place in four states.

The first issue of the company newsletter, *Hartland Courier,* arrived with the "Chart the Course" Polo Pony. Each year for six years, 2001-2006, the annual club membership included a 9" or 11" series model, a lapel pin or other small item, and a newsletter subscription. Club members also had the chance to buy or win certain models that were not available otherwise.

By late spring of 2001, new models were being produced in rapid succession. As I wrote in *Horsing Around Magazine*, November-December 2001, "Since my May/June column, two or three new Hartland models have appeared each month, only to 'disappear' again. Most models have been editions of 50, and have been quickly removed from the company website when they sell out. There's no printed catalog, and the models are not seen in stores. To avoid missing new models, it is necessary to check the website, www.hartlandcollectibles.com, or call, toll free ... at least once per month."

That was true when the article was submitted, but by the time it was published, in November, Hartland had actually mailed its first catalog – in mid-September 2001. The catalog consisted of a single sheet, printed on both sides, folded in thirds, and included 12 models or family sets. It was black-and-white. At some point in 2001, the company stopped deleting the sold-out models from its website.

**Show Special Run Model Debuts.** In May 2001, the first show special model, "Silver Sultan," an 11" Arabian, debuted. Show special models were available only from participating shows. Shows that joined the company's show producer's club, paid a fee, and complied with various rules were eligible to buy a minimum of 24 of the show special model, and sell it at the event or afterward. The shows had to pay for the models in advance. (Shows selling the specials were not announced at the company website until after payment was received. So, there was seldom much advance notice to collectors.) There was a maximum price the shows were allowed to charge; for "Silver Sultan," it was $45. In 2002, there were two show specials: "Copper King," for shows held in January through May; and "Bronze Ruler," for shows held in July through December. June shows could choose either model. Both were pinto 11" Arabians. In 2003, "Black Tie," a Polo Pony, was the show special. That was the last year of show special models. The member shows of the show producers club were, in effect, distributors. That was rare because the company usually sold directly to consumers, either on its own or through HPG.

**Test Colors Sold on eBay.** Also in May 2001, the first test color (unique) model by Hartland 2000 was sold. It was a pearled, dapple grey Polo Pony sold through an eBay auction that ended May 28, 2001. The company sold test colors on eBay from time to time. However, for production models, eBay was not the venue of choice. Only two production runs were sold exclusively on eBay, both in 2001. Mostly, the production models were sold from the company website, from its newsletter, and by mail order. That included the holiday special run models, which were issued for five years in a row (2001-2005).

**Jamboree Special Models.** In June 2001, Hartland special run models were available at the Jamboree. Hartland's horse director, Sheryl Leisure, had held the annual model horse convention as part of her business, Horse-Power Graphics, for nine years before the first models of Hartland Collectibles (2000) were released in 2001. Jamboree included model sales, a model horse show or two, and other activities such as a dinner and workshops.

**Model Horse Events.** Dozens of model horse shows, hosted by individual hobbyists, model horse clubs, 4-H clubs, or toy stores, take place around the country each year, but Jamboree was the largest model horse event in the country that was not hosted by a model horse manufacturer. Breyer's big event is BreyerFest, and the Peter Stone Company holds Equilocity and other events. However, no Hartland horse manufacturer, per se, has ever hosted its own event. Attendance at BreyerFest is in the thousands; at Jamboree and Equilocity, in the hundreds; and at most model horse shows, in the dozens. Model horse shows are also held in Europe and Australia.

In 2001 through 2004, the Jamboree included a pre-ticketed, special run model by Hartland 2000 that could be purchased only through HPG. To be sure to get one, a collector had to pay for the model in advance, and pick it up at the event – either in person or through a friend or hired agent with ticket in hand. However, an extra quantity of the ticketed, special run model was (usually) made, and when that was the case, the model was also sold at the event to people who had not pre-ordered. If models still remained afterward, they could be bought by mail, from HPG, either from its website, or in response to postal-mailings to its customer list.

Jamboree also sold the year's show (producer's club) special run model: "Silver Sultan" in 2001, "Bronze Ruler" in 2002, etc. That's because the Jamboree's main (open) model horse show, Jamboree Summer Challenge, was a member of the Hartland show producer's club. (There was also a novice model horse show held at Jamboree.)

There were also surprise special runs by Hartland 2000 sold at Jamboree, by the Hartland company itself. The models that did not sell out on Jamboree weekend were added to Hartland 2000's website, and sold from there. The Jamboree dates were: June 22-23, 2001; August 16-18, 2002; October 24-26, 2003; and October 8-10, 2004. The 2004 Jamboree was the 13th and last, but Sheryl has held similar, but smaller model horse events since then.

**Hartland Office and Communications.** From November 2001-December 2002, the Hartland company address was still in Redlands, California, but the name on the receipt had changed to "Hartland Horses," and the billing was now going through Horse-Power Graphics. A call to Hartland Horses in mid-November 2001 was answered by "Louise."

From about fall of 2001 on, each purchase of a Hartland 2000 model was (usually) accompanied by the latest catalog sheet and price list. Most new models were first announced in the company newsletter though. The newsletter was supposed to be quarterly, but tended to congregate in the second half of the year because extra models were released before and around the holiday shopping season. When new models came out in between issues, they were announced by a postcard or an email directing buyers to the company website. A person did not have to have Internet access to collect the models, but it did help.

On February 7-10, 2002, Hartland Horses sold models from a booth at Equine Affaire, a real-horse expo, at the Fairplex in Pomona, California. (Three or four Equine Affaire expos – see www.equineaffair.com – are held around the country each year; if Hartland participated after 2002, it was not publicized.)

**First Big Test Color Sale.** In May 2002, the company held a large sale of test color (unique) models via a paper catalog that was mailed to customers. It was illustrated with color photos. Some of the models were sold by silent auction, with a minimum bid of $200-$250; and others were sold at fixed prices, ranging from $75 for a 7" series foal to $195 for a 9" series adult.

Then, on January 1, 2003, the address for "Hartland Horses" changed to 34428 Yucaipa Blvd. #E119, Yucaipa, California. That was the long-time address of Horse-Power Graphics. HPG was now handling all of the orders, and the customer service person for Hartland was "Lynn." Her first name, only, appeared on communications from Hartland Horses.

In July of 2003, a resin model of Seabiscuit, the famous race horse, was available from numerous distributors on eBay. Treating it as a sports statue, the Hartland ownership was apparently connected with arranging its design and manufacture overseas (in China), and distribution. HPG was (evidently) not consulted. The resin Seabiscuit was not sold from the Hartland website, nor did it appear in the newsletter or catalog sheets. It really was a Hartland Collectibles product, though. The box reads, "It is only fitting that Hartland Collectibles, widely regarded as the first company to manufacture horse replicas over 50 years ago, be the ones to produce this special Seabiscuit statue." (That's a slight re-write of history since the original Hartland company, in the 1940s, was Hartland Plastics, Inc., not Hartland Collectibles, L.L.C., which did not exist until 2000.)

In July 2004, Hartland Collectibles, L.L.C., announced that it was "now doing business as Hartland Sports and Hartland Westerns. Same ownership, just divided into two divisions with new names and logos." The receipts enclosed with the model horses continued to be from "Hartland Horses."

A new development later in 2004 was that Hartland tried a different painting service for five horse models released in December 2004. Saying that those models had small paint runs or other tiny flaws, the company called them "sample models," and sold them for only $19.99. This painting service was evidently not asked back, but it did paint some of the most complicated eyes. Normally, a sample is a prototype, not an edition.

By late in 2005, Hartland Horses moved with HPG to Mariposa, California, in the northern half of the state. Receipts from December 2005-December 2007 were from: "Hartland Horses / Distributed by Horse-Power Graphics, Inc." The address was a post office box in Mariposa, California. When a phone call was placed to Hartland in Mariposa, it rang in Sheryl Leisure's house. By late 2005, HPG was the sole distributor of Hartland 2000 horses in plastic, in addition to its role in designing the model colors.

**Second Big Test Color Sale.** The second large sale of test color (unique) models was announced in December 2005. The prices were $25-$45 for a 7" series horse, $65-$75 for 9" series adults, $45 for a 9" series Weanling Foal, and $95 for an 11" series Arabian. The description said that the models had been "painted from the year 2000 to present," and that, "Test models are generally painted on models of non-salable quality, so rough seams are common." I think the company was being pickier than most collectors. Hartland models, back to the 1940s, often have discernable seams; personally, that has never bothered me.

December 2005 was also when the Roy Rogers and Dale Evans horse-and-rider sets were finally released. Stickers on the models read, "© 2003 Horse-Power Graphics, Inc." That may be the year in which the color scheme was designed. Hartland had taken deposits for the sets in January 2004. Then, in mid-2004, the deposits were refunded because the models' release was stalled by a legal problem. A third party that had been licensed to sell Roy Rogers items complained that Hartland was also licensed. Hartland did not say who the third party was.

From November 2005-February 2007, the Hartland Horses price sheets listed back issues of the *Hartland Courier* for 25 cents each.

In summer of 2006, Hartland Collectibles, L.L.C., moved from Shreveport, Louisiana, to the Cleveland, Ohio area. It was still owned by the same group of investors, but had new management. The sports half of the company website changed entirely, but the horse and western half remained the same.

**A Horse in December 2007.** A Semi-Rearing chocolate bay with saddle, announced in late December 2007, was the last horse released as of early 2012. A run of only 20, it sold out quickly.

Models painted from 2001-2007 were still being sold in the company's eBay store (hartland_ horses) until August of 2008. By then, the company's website for horse and western models, www.hartlandcollectibles.com, was mostly deactivated. Only one page of it was visible, and orders could no longer be placed.

I found a bankruptcy date of August 7, 2008, for Hartland Collectibles, L.L.C., at *Business-Week* online. An undated, online article in *Tuff Stuff Magazine* (probably from late August 2008) said that Ohio-based Hartland Collectibles, L.L.C., "was being dissolved through Chapter 7 bankruptcy liquidation," to cover debts through 2007. When it filed for bankruptcy protection, it listed assets of $32,000 and liabilities in excess of $721,000, the article said. However, sports figurines were produced in 2008 under a different corporate name, Hartland of Ohio, L.L.C., and that production was not part of the bankruptcy. (Since then, Hartland of Ohio, L.L.C., of Mentor, Ohio, has released additional sports and other figurines, including figurines of Barack Obama, the 44th President of the United States. New products for 2012 can be ordered at www.hartlandllc.com.

The financial difficulties are, no doubt, the main reason for no new releases of model horses since December 2007.

**Models Still for Sale on eBay.** Since the Hartland Collectibles (2000) website and its eBay store became inactive in 2008, the remaining horse and western items produced by Hartland 2000 and leftover Hartland stock from the 1990s have been for sale in HPG's eBay store (horse_power_graphics_inc). In June 2011, in an eBay auction for unpainted Hartland baseball bat boy statues, HPG (Sheryl Leisure) stated, "These and much of the other old Steven and Hartland Westerns stock was turned over to the studio in lieu of bills owed." (The "studio" was Sheryl's; she painted the prototypes for the horse and western models by Hartland 2000.)

The Hartland Collectibles website, www.hartlandcollectibles.com, had not functioned for at least two-and-one-half years since August 2008, but when I typed that address in June 2011, it took me to a page at www.modelhorsejamboree.com that showed the Hartland Roy Rogers and Dale Evans rider sets for sale. The front page of the Jamboree website, which is owned by HPG, said it had been updated in May 2011. That was after I had contacted Sheryl by email about this book in April 2011. Sheryl she said she regretted that she did not have easy access to her Hartland company records since they were stored away with her collection. She answered a few questions, and that was it. Other questions I had on issues small and large – mostly things she could answer without looking in records, and including the most obvious question, "Who owns the horse and western part of Hartland now?" – were unanswered. She declined my invitation to read over this chapter of the book.

Early in 2012, Sheryl was kind enough to answer a few more questions, and she summarized her involvement with Hartland. She said that she worked for Hartland on a contract basis, as a consultant, and then also became a distributor. She said that Hartland put her business name, Horse-Power Graphics, Inc., on the Roy Rogers and Dale Evans boxes (and on the stickers on those models) because she handled the ordering, model paint-scheme design and package design, and the distribution and sales of those sets.

She designed the new Hartland model horse colors, but the painting of the entire runs was usually contracted out elsewhere by Hartland. There were, however, some small runs she painted herself. She said those could be considered "manufactured by Horse Power Graphics, Inc., under license to Hartland Collectibles, L.L.C."

If I understood correctly, those were the six models released from May 2006 through January 2007: the Liver Chestnut 9" Five-Gaiter; the Red-Chestnut Overo-Pinto Polo Pony; the Bay Roan and Chestnut Roan Tennessee Walker families; "Viceroy," the Dappled Buckskin 11" Quarter Horse; and "Volant," the Dappled Palomino Polo Pony with green wraps. "Viceroy" and "Volant" were signed by Sheryl, but it's nice to know that these other models were also painted by her. (Sheryl said that she did not paint the run of "Coco," the Semi-Rearing horse from December 2007.)

In June 2011, the models still available from HPG on eBay included the Roy Rogers and Dale Evans rider sets, blue Roy Rogers saddle (not part of a set), Indian rider accessories, and some 7" Tennessee Walkers, all painted by Hartland 2000. Steven stock from the 1990s included the grazing Arabian mare (rusty grey, new in box), 1994 Chief Thunderbird rider and saddle (painted, but no horse or small parts), George Washington saddles (painted and unpainted), Chief Thunderbird and his accessories (unpainted and unassembled, and either on or off the sprue); and Gen. Lee, Gen. Washington, and Dale Evans riders and accessories only (unpainted). Early in 2012, most were still available. For some rider-series items that are more scarce than production runs, see Chapter 18.

**Things That Did Not Happen.** Every previous era of Hartland has experienced plans that did not reach fulfillment, and Hartland 2000 is no exception. In the fall of 2002, Hartland announced its plan (in *Horsing Around Magazine*, November/December 2002) to issue a resin version of the Lady Jewel and Jade Arabian models with modifications and additional details added by Sheryl Leisure. (Lady

Jewel and Jade were sculpted by Kathleen Moody for Steven Manufacturing in 1988.) The names of the resins were to have been "Lady Diamond" and "Gemstone." They were never released.

In September 2004, Hartland announced that the Brave Eagle rider set would return early in 2005. In December 2004, the date was pushed back to summer 2005. Brave Eagle, an Indian chief, was the lead character in a television show of the same name that aired from 1955-1956. The program was owned by Roy Rogers, and the original Hartland Brave Eagle production dates were, by my estimate, 1957-1960 (*Hartland Horsemen*, p. 133). Hartland's Brave Eagle has not yet returned, but the Roy Rogers and Dale Evans sets did.

Some entirely new horse shapes in resin were advertised, but then not produced by Hartland. A new sculpture in resin, "Walter" the trotting Quarter Horse gelding, was slated to be available at the Jamboree in June 2001 along with a copy of the paperbound book, *Walter Spills the Oats,* by Don Blazer. The model wasn't ready, but a sample was displayed, and $10 deposits were taken. Hartland also displayed two other resin samples at Jamboree: "SweetHart," a jogging stock horse mare, and "Harts Afire," a trotting Arabian mare. All three resins were advertised in the company's Spring 2001 newsletter. Hartland also advertised "SweetHart" and "Harts Afire" in *The Hobby Horse News,* August/September 2001. In *Horsing Around Magazine*, it advertised "SweetHart" in May/June 2001 and "Walter" and "SweetHart" in September/October 2001. Finally, in June of 2002, Hartland said that the resins would never be available. Those who had put down the $10 deposit received the *Walter* book.

The sculptor, Sheryl Leisure, produced the resins herself, and sold them under the name, Horse Country Collectibles, a division of Horse-Power Graphics, Inc. By November of 1992, "Walter" (a black bay), and "Change of Hart" (the "Walter" mold in four other colors) were available along with a cantering Friesian ("Hart of Thor") in black and four other colors. The resins were mass produced in China with excellent workmanship, and were priced at $65 each. (The prices have been lowered significantly since then, making them an even better value for the price.) In 2005, sculptor Leisure produced the Arabian, "Harts Afire," and a reining horse, "Turnin' Harts." Those two were cast in smaller quantities by a business called Resins by Randy (Buckler) at www.resinsbyrandy.com. They were available painted or unpainted, at prices starting at $125.

**What's in A Name?** Although they were not Hartland products, the Horse Country resins still carried names with "Hart" in them: "Change of Hart," "Hart of Thor," "Harts Afire," and "Turnin' Harts." The chance to make a play on words was not lost on the Hartland production models, either. Four of them had "Hart" or "Heart" in their name: "Steadfast Hart" (2003), "Heart of Gold" (2004), "Hart of Blues" (2005), and "Lion Hart" (2006).

("Heart of Gold" was probably intended to be "Hart of Gold," but the decorative tag that came with the model was printed with the word "Heart," and the company stuck with that spelling until after the model was being sold without the tag.)

The holiday models usually had Christmas-related names: "Sugarplum," "Lump of Coal," "Noel," and "Glisten" in 2001; "Snow Angel" in 2002; "Frankincense" in 2003; and "Christmas Chocolates" in 2005. Three models had names connected with warm beverages: "Cuppa Joe" and "Java Joe" in 2005 and "Cocoa" in 2007. The two horses in the Noble Horse Series had French-derived names starting with "v": "Volant" (2007), which means "nimble," and "Viceroy" (2006). Three show specials also had names evoking royalty: "Silver Sultan," "Copper King," and "Bronze Ruler." After all, they were part of the "Regal" (11") series.

Other model names touched on a range of themes. "Chart the Course" was the first model from Hartland 2000. "Tom-Tom" was a pinto, a color favored by American Indians. "Wave the Banner," which came out five months after 9-11, struck a patriotic note. "Jammer Time" was a Jamboree special. A model in a pinto pattern resembling formal wear was "Black Tie," a show special. A buckskin model was "Buckeye." A metallic copper horse was "Cinn," perhaps, in reference to cinnamon, a color not far from copper. Two gift run horses had names, too: "Ascot" and "Dapper" were natty-looking 7" Tennessee Walker stallions.

**Test Display in 2009; Sample Sale in 2010.** Hartland test color models were displayed at the Jamboree Winter Challenge Model Horse Show, held by Sheryl Leisure on February 28-March 1, 2009, at the Mariposa County Fairgrounds, Mariposa, California. The test colors were 5" and

7" series family-set horses. They were not for sale, but at least one collector photographed them.

In November 2010, Sheryl sold Hartland "sample" models on eBay. These were models that Steven Manufacturing had used as working samples in painting the late 1980s and early 1990s Hartland production runs. The samples included breed-series horses, rider series horses, and riders. Only a few of the samples were different from the production colors. Most were slightly beat up from frequent handling, and most auctioned for under $25 as souvenirs.

These factory worn "samples" contrast with the "test" (unique) models from Hartland 2000. The Hartland 2000 tests were treated as the valuable models they proved to be when sold to eager collectors.

The 1980s-1990s samples will be described in a later book. Additional history of Hartland models released after 2000 is found in the descriptions of individual models and sets.

## Small Editions, Great Details

History is more than a collection of facts. Analysis and commentary can shed more light.

The Hartland models since 2000 are remarkable for their small editions, and the painstaking work that went into them. Small editions aimed at collectors were the company's objective. A quick sell-out was to the company's financial advantage, but being more flexible about the edition sizes could have benefitted both the company and the collectors. Especially in the first year the company was in business, popular models sometimes sold out in a few weeks, leaving a share of the collectors empty-handed. The tactic of turning people away, instead of making them happy, seemed like a bad idea. Simply making a few more models to accommodate the orders would have been an easy solution.

**Small Runs.** The production models were relatively difficult to acquire. The models were marketed mainly through the company's website and its postal-mailings; the models were not sold in retail stores or by multi-brand mail-order houses.

For most models, the editions were limited, and often, the run size was only 50. That was true for many of the regular-run models. Special-run models, which were models for special occasions and/or that could only be purchased under special circumstances, added another, or different, layer of difficulty. Holiday special models, while usually unlimited in quantity, could only be ordered during a narrow window of time, such as 6-12 weeks. The club special models were tied to an annual membership, which, in effect, raised their price.

Show special models had to be bought at the event or ordered afterwards from accommodating show hostesses. The event known as "Jamboree," also had its own special run; a ticket for the model had to be purchased in advance, and the model had to be picked up in person. (In some years, it was possible to buy the Jamboree special without an advance-purchase ticket, but the price was higher.)

**Special Run Model Horses or Horse Sets**

| Type of Special Run | Qty. of Model Editions |
|---|---|
| Club Models | 7 |
| Holiday Models | 6 |
| Small Holiday Model w/ larger model | 2 |
| eBay Specials (not Holiday) | 1 |
| Show Specials | 4 |
| Noble Horse Series and "Cocoa" | 3 |
| Jamboree Ticketed Specials | 4 |
| Surprise Jamboree Specials that sold out at the event | 4 |
| Sold through Drawing in Newsletter (Not including Noble Horse series) | 4 |
| **TOTAL SPECIAL EDITIONS** | **35** |

**Special Runs** – Among the production models by Hartland in 2001-2007, 35 different model horses (or horse sets) were sold as special runs (as opposed to regular runs). There were nine categories of special runs: club models, holiday models, etc.

Another type of special run was models sold by the Hartland company at a booth at the Jamboree. The identity of the models was a surprise, they could not be pre-ordered, and they sometimes sold out within hours at the event. Then, there were production models that could only be purchased by the group of winners among those who entered a random drawing through the company newsletter, which was only mailed to those who had joined the company's "club."

There were a lot of special runs in proportion to the regular runs, and a lot of hoops to jump through to acquire them. Because I made a concerted effect to monitor the Hartland model releases, and had help in buying models from distant events, I missed almost none of the production models. However, I know of collectors who could not pay such close attention. They were disappointed to miss out on some of the models they would have bought.

**Note to model horse manufacturers / distributors:**
Keep it simple! Make it easy for us to buy your products! Make collectors feel included, not excluded. Also, customer service really matters.

### Total Production Models
(by Hartland in Plastic), 2001-2007

| Type of Model Edition | Qty. of Model Editions |
|---|---|
| Special Runs: Horses (lone or sets) | 35 |
| Regular Runs: Horses (lone or sets) | 47 |
| TOTAL Horses / Horse Sets | 82 |
| Accessories (lone or sets) | 12 |
| **GRAND TOTAL: Horses and Accessories** | 94 |

**Total Production** – Of the horse items in the catalog, 35 were special runs compared to 47 regular runs. Adding in the accessory items brings the grand total to 94. Also, some of the horse items were actually multi-piece sets. Adding in the extra12 members from the six 7" Tennessee Walker families, and the 30 extra horses from Tinymite groups, the total number of production-run model horses was 82 + 42 = 124.

**Excellent Workmanship.** The positive side of the small editions is that it probably made it more feasible for the company to put the amount of detail into the models that it did, and still keep the prices reasonable. These Hartland horses included a greater percentage of dappled and/or pearled colors, and more spotted horses (pintos and appaloosas) than any previous Hartland company. In addition, many of the models had three colors, or even four colors, painted into their eyes.

This group of Hartland horses since 2000 comes close to custom-painted quality, rather than typical mass production in which thousands of models are produced in each color. Because of their artistic sculpture, it's hard to make a Hartland model look bad, but this company really made them look good.

## Original Prices and Values

The Hartland models since 2000 started out at prices just a little higher than Breyer horses and less than one-half the price of many Peter Stone horses of comparable size. The Hartland 11" series horses, which are of the same scale as the majority of Breyer and Peter Stone horses, were often $36 while the 9" series horses were $32, and three-piece horse families, 7" series, were $45.

In January 2003, however, the prices for most models were lowered, brought more closely in line with Breyer prices. For regular run models, the 11" series went down to $32; the 9" series to $29 or $30; and the three-piece, 7" series families became $38. At either the original prices or the 2003 prices, the Hartlands were a good value for the price.

Values of collectibles in general reached a peak in about 1998-2002, fueled by the mid-1990s prosperity and eBay's honeymoon period, but have coasted downward ever since, especially since the recession that began in December 2007. Common collectible items of any brand or type now sell for maybe about half of what they did at the peak. (The values in my two earlier Hartland books were deliberately kept "sane," and so are now not far off the mark in the current, cooled-down market.)

Vintage, rare items with a strong following still sell well on eBay, but the Hartland horses since 2000, while scarce, are too recent to be truly appreciated, and do not have a large following (yet). Nobody made horse-and-rider sets like Hartland did in the 1950s and 1960s, but the regular (breed-series) Hartland model horses made from the 1960s to present, are just one among many brands of popular horse figurines in plastic, china, and resin.

High-quality commercial resins sell for $35-$100, and hundreds of limited-edition artist resins are often $200 or more each. A horse collector's dollar has many directions to be pulled in, and stretched thin, these days. One resin purchase can use up the money that might have been spent on six or more plastic horses before resin horses were an option.

The model horse market has changed a lot since the 1950s-1960s heyday for Hartland Plastics, Inc., and the Hartland revival in the 1980s and 1990s, but beautiful plastic horses will always find a home.

# BIBLIOGRAPHY AND RESOURCES

**Advertising and news from the Hartland company in model horse publications.** The Hartland company submitted ads and news releases to *The Hobby Horse News* and *Horsing Around Magazine*. The articles appeared without bylines. Also noted are Horse-Power Graphics, Inc.'s Jamboree ads regarding Hartland models ("J.ad") and its ads for resins from Horse Country Collectibles ("R.ad").

In: *The Hobby Horse News* (while published by SAG Communications, Greenville, South Carolina):


Issue 73 – December 2000/Jan. 2001: news, p.20
Issue 74 – Feb./ March 2001: news, p. 6-8; ad, p.22
Issue 75 – April/May 2001: news, p. 22; ad, p.45
Issue 77 – Aug./Sept. 2001: ad (resins "SweetHart" and "Harts Afire"), p.49
Issue 78 – Oct./November 2001: ad (resins "Walter" and "SweetHart"), p.51
Issue 79 – December 2001/Jan. 2002: ad, p.30
Issue 80 – Feb./March 2002: ad, p.38; J.ad, p.39
Issue 81 – April/May 2002: ad, p. 50


*The Hobby Horse News* (1988-2004), was produced by Paula and David Hecker, Florida, until Tina Ferro became editor in 2000. Shown is Issue 79, December 2001/January 2002. Model horse publications produced by hobbyists date back to 1969, and include *The Model Horse Shower's Journal, American Model Horse Collector's Digest* (later *The Equine Miniaturist*), *High Steppers Review*, and *The Model Horse Gazette*, to name a few. The journals took hold in the 1970s, and flourished in the 1980s and 1990s.

In: *Horsing Around Magazine* (published by Horsing Around, England; www.horsingaround.com):


Issue 44 – March/April 2001: ad (7" TW St.),p.11
Issue 45 – May/June 2001: J.ad, p.14; ad (resin "SweetHart" and plastic models), p.37
Issue 47 – Sept./Oct. 2001: ad (resins "Walter," "SweetHart," and a plastic model), p.7
Issue 48 – November/December 2001: ad, p.10
Issue 49 – Jan./Feb. 2002: ad, p.24
Issue 50 – March/April 2002: ad, p.8; J.ad, p.9
Issue 51 – May/June 2002: news, p.12; ad, p.24
Issue 52 – July/Aug. 2002: ad, p.32; J.ad, p.37
Issue 53 – Sept./Oct. 2002: news, p.8-9; ad, p.22; R.ad, p.37
Issue 54 – November/December 2002: news, p.10; R.ad, p. 41
Issue 55 – Jan./Feb. 2003: news, p.15; ad, p.20; R.ad,p.25
Issue 56 – March/April 2003: ad, p.41
Issue 57 – May/June 2003: J.ad, p.5
Issue 58 – July/Aug. 2003: ad, p.29; R.ad, p.44
Issue 59 – Sept./Oct. 2003: ad, p.28; J.ad, p. 48
Issue 60 – November/December 2003: ad (Roy Rogers & Dale Evans), p.48
Issue 61 – Jan./Feb. 2004: ad, p.8
Issue 62 – March/April 2004: R.ad, p.6
Issue 64 – July/Aug. 2004: ad, p.28; J.ad, p.37
Issue 65 – Sept./Oct. 2004: J.ad, p.8; ad, p.37
Issue 66 – November/December 2004: R.ad, p.28; ad, p.41
Issue 67 – Jan./Feb. 2005: R.ad, p.18; ad, p.25
Issue 68 – March/April 2005: ad, p.36
Issue 69 – May/June 2005: R.ad, p. 2; ad, p.37
Issue 70 – July/Aug. 2005: R.ad, p.6; ad, p.25


*Horsing Around (International Equine Sculpture) Magazine* (1994-2005) was published in England by Horsing Around. The business, owned by Mark and Vanessa Crawley, still produces limited edition china and resin horses and related products. Except for *Just About Horses*, the Breyer company's magazine, *Horsing Around* was the first model horse periodical to print at least partly in color. American advertisers and subscribers flocked to it. Shown is Issue 45, May/June 2001.

**Miscellaneous Sources.** Other sources besides those mentioned in the text, are:

Articles / Photos in *Hartland Model Equestrian News* (published by Gail Fitch):


Barnwell, Lauri. "Victorian Village Live – Show Report," Jan.-Feb. 2002, p. 52-53
Meathrill, Suzy. Photos of resins, "SweetHart" and "Harts Afire," May-June 2001, p.18
Pramuka, Pamela. Photo of flecked / flea-bit grey Polo Ponies, Sept.-Oct. 2004, p. 65

## On the Internet:

Teller, Melanie. Photos of test-color 5" and 7" family-series models displayed in February 2009; a link to the pictures was in her March 10, 2009 message to hartlandmodelhorses at www.yahoo.com.

## Published works by the author:

### Articles about Hartland by Gail Fitch in *Horsing Around Magazine,* published by Horsing Around, England:

"Horses, Models, and Admirers" (part 1), Issue 45, May/June 2001, p.11-12
"Horses, Models, and Admirers" (part 2), Issue 48, November/December 2001, p. 14-15
"Galloping Across the Page," Issue 51, May/June 2002, p. 38

### Articles by Gail Fitch in 12 issues of *Hartland Model Equestrian News*, published by Gail Fitch:

"Hartland Models to Return," p. 1, November-December 2000
"Hartland Plans Clubs, Direct Sales," p.2, November-December 2000
"First Hartland Model of the New Century Arrives," "Jamboree Features Three Models," and "Prices Announced for Hartland Models," p.10, March-April 2001
"New Hartland Models Benefit from Pearl White," p.12, March-April 2001
"Five, Surprise Hartland Models Sold at Jamboree," p. 17-18, May-June 2001
"Three Hartland Resins Planned," and "Other Hartlands at Jamboree," p. 18, May-June 2001
"Models Left Over From Jamboree," and "Silver Sultan Schedule Grows," p. 20, May-June 2001
"Book About 'Hartland' Model" [Walter], p. 22, May-June 2001
"Hartland Special in eBay Store," p. 25, July-Aug.2001
"Hartland Size and Color Terms; Pearled Colors Compared," p. 26, July-Aug. 2001
"The 2001 Hartland Models Have Numbers" [table of models], p. 27, July-Aug. 2001
"Four Shows This Fall Feature 'Silver Sultan,'" p. 28, July-Aug. 2001
"Pearled Finishes Enhance Three Out of Five June Specials," p. 29, July-Aug. 2001
"Hartland Models for July: Mostly Black," and "Buckskin Mustangs in Two Versions, p.29, July-Aug. 2001
"Hartland Pinto Five-Gaiter," p. 37, Jan.-Feb. 2002
"Summary of Model News for An Unusual Autumn," p. 40, Jan.-Feb. 2002
"2001 Hartland Models Since August: 15 Shapes of Horses" [including table], p. 41, Jan.-Feb. 2002
"Hartland Summary: Models Still Available, Made/Sold from 2001-Present" [table], p. 64, Sept.-Oct. 2004
"Twelve Images of Hartland Horses from mid-2001 to early 2002," p. 65, Sept.-Oct. 2004
"Special Hartland Sale," p. 66, Sept.-Oct. 2004
"Hartland Theme for June 2002: Gray," "Warm Colors," and "Hartland Laments" [re: resins], p. 67, Sept.-Oct. 2004
"Copper King," p. 68, Sept.-Oct. 2004
"Sculptor Sheryl Leisure to Issue Horses Including 'Walter' Resin," p. 69, Sept.-Oct. 2004
"Selected Model News," p. 70, Sept.-Oct. 2004
"Prices Lowered on Hartlands!" p. 73, Sept.-Oct. 2004
"It's Time to Accessorize..." p. 75, Sept.-Oct. 2004
"2003 Hartland Horses," and "'Hartland Seabiscuit: Yikes!" p.76, Sept.-Oct. 2004
"Model Jamboree, October 2003," p. 78, Sept.-Oct. 2004
"Hartland Tinymites," p. 80-81, Sept.-Oct. 2004
"New Hartland Models for 2004," p. 81, Sept.-Oct. 2004
"Hartland Requests Feedback," p. 82-83, Sept-Oct. 2004
"Hartland Update: Models (Still) Available, Made/Sold from 2001-Present" [including table], p. 92, November-December 2004
"Holiday Horse by Hartland," p. 93, November-December 2004
"Three Hartland Items at Jamboree," p. 93, November-December 2004
"2004 Hartlands: Six Sold, One 'Gift,'"p. 94, November-December 2004
"Brave Eagle Indian Set Planned," p. 94, November-December 2004
"Five 'Sample Run' Models Issued in December," p. 120-121, Jan.-Feb. 2005
"Hartland Brave Eagle Set Pushed Back Again," p. 121, Jan.-Feb. 2005
"Buckskin ('Coffee Dun') is Hartland 2005 Club Model," and "Hartland Tinymites," p. 123, Jan.-Feb. 2005
"Seabiscuit" [Hartland], p. 126, Jan.-Feb. 2005
"Hartland Roy Rogers and Dale Evans Return," p. 142, November-December 2005
"Old Stock, New Models Available from Hartland" [including table], p. 144, November-December 2005
"Hartland Favors Dun Horses," p. 145, November-December 2005
"Many New Hartlands Debut in December, January," p. 168-169, July-Aug. 2006
"New Hartland Horses in June," and "Test Color Update," p.170-171, July-Aug. 2006
"Hartland Palomino Starts 2007," and "Hartland Horses Still Available" [list], p. 188, Spring 2007
"Company Sells 1990s Old Stock," p. 189, Spring 2007
"Indian Test Color," p. 190, Spring 2007

### Articles by Gail Fitch in *Model Equestrian: Horse Collecting & Hartland News,* published by Gail Fitch:

"Hartland Accessories Released," p. 9, Fall 2007
"Hartland Mostly Inactive," p. 35, December 2008
"Unpainted Models," p. 22, November 2010

Thrilled that Hartland horses were once again being produced, the author reported on them at her website and published in depth in *Hartland Model Equestrian News* (12 issues, late 2000 through summer 2007), and since then, under a new title: *Model Equestrian: Horse Collecting & Hartland News.* Shown is the January-February 2002 issue. Jennifer Lang customized the Hartland horses on the cover.

## Books (the most recent titles) by Gail Fitch:

*Hartland Horses: New Model Horses Since 2000* (first edition, 2010)
*Hartland Horses and Dogs* (Schiffer Publishing, 2000)
*Hartland Horsemen* (Schiffer Publishing, 1999)

## Website by Gail Fitch:

Hartland Model Equestrian, www.hartlandhorsemen.com (since July 2000)

**History.** The list of article titles is, in itself, a short history of Hartland since 2000.

# ABOUT THE AUTHOR

My parents both like dogs, and all four of us children like animals, but I fell in love with horses (or horse likenesses) at an early age. As a preschooler, I saw horses on television, and enjoyed some hollow-bottomed, plastic toy horses. I had a white one I named "Pony Marshmallow." As to the real thing, my earliest memory is of seeing two white ponies close to the fence near a switchback in rural Tennessee in July of 1956, when we were returning to Milwaukee. (My father's an aerospace engineer, and his company had sent him [us] to Savannah, Georgia, for a year.)

In 1958, I was awed by Hagen-Renaker mini horses at a gift shop owned by relatives, Lester and Theodora Lueloff, in Fond du Lac, Wisconsin. For good behavior during the visit, my mother bought my brother and me each a horse, and Great Aunt Teddy threw in the matching foals!

A young Gail Fitch, shown with her father, Lawrence, is enthralled by a wooden horse likeness on wheels.

They were kept in the china cabinet, and I was told to not touch. I couldn't resist, and in my nervousness at doing a forbidden thing, dropped my favorite. (It was repaired, but a foot was lost.)

In 1959, I went to my first horse show, with my father. It was all-Arabian, and I still have the program. I'll never forget the cutting horses, and the trail class. In school the next week, I drew a crayon portrait of the horse that impressed me most: a dark chestnut that had been ridden under English tack. I actually drew two pictures of it because a substitute teacher asked me to make one for her.

Playing horses got me in trouble in 1960. On a bus returning from a Girl Scout outing, I sat at the front of my seat, pretending to be on a bouncing horse. When the bus stopped abruptly, I chipped a front tooth on a metal part of the seat ahead of mine. Several dental procedures followed over the next year. After the last session, with my mouth still numb, my mom took me to the toy department in the basement of Grant's (a dime store / department store), and said I could have anything I wanted. I would have picked a red plastic moose, but then I saw a horse: an Aurora Black Fury model kit. So, there was a silver lining.

In 1961, I helped my father's mother pick out two Hartland mini rider sets at the Rexall drug store on Greenfield Avenue, near Layton Boulevard, where she lived. One was for me, and one was for John, the older of my two younger brothers. In fall of 1962, my maternal grandmother returned from a trip, and surprised me with the gift of an elegant, china horse made in Japan.

In the 1960s, I received other horse items – books, toys, plastic models, a china horse head wall plaque, a stuffed horse ("Fluffy") – as gifts. (I mention this not to boast, but to show my appreciation to the gift givers.) A Sunday school teacher, Dorothy Mann, gave me *Horses: A Golden Stamp Book* for no reason at all; it wasn't Christmas or my birthday or anything.

I also started horseback riding. Some Christmases, my parents gave me a homemade coupon good for five rides! I took group riding lessons, English style, at Joy Farm for a few years and took a one-hour, western trail ride about once per year, usually at Parkway Stables, near Whitnall Park. (It amazes me that riding stables were still so close to Milwaukee, which, at that time, was about the 10th most populous U.S. city.) Joy Farm's specialty was indoor polo; the stable taught balanced seat riding. Sometimes, there were thirty people – mostly girls from 8-18, riding in pairs around the ring! It could get dusty!

Joy Farm had its own bus service, with a stop about five blocks from my house, which was why it was easy for me to go riding on Saturdays without parental assistance. I paid for riding with allowance money (for doing yard work and household duties), and my earnings from taking neighborhood children to school. The school, Maryland Avenue School, which was K-8, was eight blocks away, and most of the route was along a very busy street, North Avenue. The first semester of taking a neighbor, Marion Koehler, to school, my mother stockpiled my earnings (25 cents per week), and surprised me by using it to buy a book for me, Marguerite Henry's *Album of Horses*, for about $3.50. Later, another neighbor paid $1.50 per week, which went pretty far since riding was $2 per hour.

As part of my plan to eventually have a horse, I bought some inexpensive riding attire from a mail order catalog, *Beckwith's of Boston,* in summer 1963, with my parents' permission. (I paid *them*, and they wrote the check.) Then, my father took me on two shopping expeditions at my request. I bought a bit and hoof pick at The Milwaukee Saddle Shop when it was going out of business in December 1963, and bought a bridle and grooming implements at Elsner's Saddlery the following spring. (At Joy Farm, a number of the riders brought their own bridles, but hardly anyone brought a grooming kit. When I asked permission to groom a horse, the stable man must have been quietly amused.)

I only rode in good weather, but read year 'round. I could walk to two public libraries, and we occasionally went to the big, downtown library. I read about horses: both fiction and non-fiction. I also scanned the newspaper for horse articles or pictures, and made scrapbooks of the clippings.

For my birthday, in February 1963, I received a subscription to *Horse Lover's Magazine*. That was probably my mother's idea. Then, in September 1963, I started a monthly digest of horse news and features. I did 12 issues. It included original writings and drawings, and articles I hand-copied from the newspaper. I made only one copy, for myself, so nobody's copyrights were being violated, although of course I was not aware of such things then. In 1964, I selected a subscription to *The Horsemen's Advisor*, which turned out to be a Saddlebred and show horse publication from Iowa, and then two years of *Arabian Horse World*.

I started holding little horse shows for my toy and model horses in fall of 1963. I cut one-inch rosettes from colored ads in the picture sections of the Sunday newspaper. For one show, I typed a program on my parents' typewriter, using the red half of the ribbon, so the black-inked half would not wear out! (It was a 1927 Royal typewriter with glass sides that my parents had bought second-hand in 1949, when my father was studying mechanical engineering at Marquette University. He was required to submit typewritten metallurgy lab reports. My mother typed them. Much later, I used the typewriter for my earliest Hartland books.)

In 1964, a fellow horse lover, Susan Sawyer, brought a Red Bird Sales Co. catalog of Breyer, Hartland, and china horses to school. It had been advertised in *The American Girl* (the Girl Scout magazine that was published from 1917-1979). I sent for the catalog, too, and daydreamed about what horses to get, but didn't buy any of them then. Instead, on July 3, 1964, I bought two Breyers and a Japan china at F.W. Woolworth Co. downtown. The catalog was my most prized possession, though.

I made English saddles and bridles for some of my horse statues, and in 1965, I built a stable with cardboard walls and stalls and a plywood floor. It was 30" x 19" and had room for about 20 min-sized horses. That was also the year I made a 9" series western saddle out of thin, imitation leather, using my Hartland Wyatt Earp saddle to draw the pattern pieces.

In 1966, my parents allowed me to join the American Horse Shows Association. (Again, I paid them, and they wrote the check.) I joined so I could get the rule book and read about all the divisions and classes and what the different classes were judged on. In those days, the AHSA covered nearly every breed; that was before the Quarter Horse people and others split off to do things their own way.

During Sunday sermons in the 1960s, I would sketch horse heads and tack in the margins of the church bulletin. Once, by invitation, I brought my bridle and grooming implements to Sunday School. I probably talked about horse equipment with the enthusiasm of a museum curator discussing rare artifacts. The teacher, Mrs. Bartelt, could only hope in vain that my fervor would spill over to the curriculum.

(That reminds me: There was a Hartland Large Champ Horse and Cowboy on the fireplace mantel in the basement of the Sunday school building in 1963. The church was, by then, in a bad neighborhood, and things would disappear. By 1969, the horse and rider were gone.)

In (public) school, I gave bridle demonstrations twice: in seventh grade and tenth grade. My fourth grade teacher, Miss Elaine Miller, included horse books in the order for the class library, and announced that I could be the first to read *Black Gold*. She also ordered *Diving Horse*.

From about fourth through eighth grades, I usually found a way to make horses the subject of just about any art or creative writing assignment. That once drew the attention of the school district's art supervisor; he told me that I should draw other things.

My fifth-grade teacher, Amy Boening, encouraged me by having me field the class' questions about horses for about a half-hour at the front of the room one day. She also had me read a creative writing piece – a breathless account of a fox hunt – aloud to the class. I've never been on a fox hunt, but it must have been pretty good, because someone stole it from the teacher's desk afterward.

More often, though, classmates gave me things, simply because they knew I was interested in horses. I became the honored recipient of: a page torn from an old dictionary, with etched images of horse breeds on it (including German Coach and French Coach horses); a playing card with a palomino horse head (somewhere, a deck is short a card); and a one-inch-high bone china pinto. (Thank you, Susie Walter, Helen Hess, and Barbara Felde.)

A sixth-grade teacher, Mr. Diek, was less tolerant. He tore up the Pegasus I'd made as Halloween art. The following year, Miss Grandy didn't mind that I'd drawn the Headless Horseman. A novel, *Blitz: The Story of a Horse*, by Hetty Burlingame Beatty (1961), evidently made a big impression on me, because for Fire Prevention Week in 1964, I drew a poster that said, "Remember the loyal fire horses." My parents kept it on a wall in the basement until we moved to the suburbs in 1966. The more I think about it, I was a pretty weird kid, and my parents were mighty understanding.

Journalism was one of my best subjects in both high school and college (but my major was sociology). After college, I worked at a publishing company, and then did research and writing in municipal government. I got very interested in model horses again late in 1974, a year after college.

Seeing Beswick (china) horses at the downtown Boston Store kicked that off. I bought a brown bay Thoroughbred. Then, I bought six Breyers in one week between the Drews' Variety Store at Point Loomis Shopping Center and F.W. Woolworth at Southgate Shopping Center. (Both shopping centers have been mostly torn down since then; one was replaced with a Walmart.)

I got connected with the organized model horse hobby in 1975, and have participated ever since. That is the year I started subscribing to model horse journals. I found out about Breyer's *Just About Horses* from an article in *The Morgan Horse* magazine in spring 1975. By mid-year of 1975, I was also subscribing to Linda Walter's *The Model Horse Shower's Journal*, which was full of sales lists and class lists for photo shows by mail.

Linda Walter's model horse publication (1969-1980) included not only full-page ads and brief articles, but also social news notes by Linda that gave *The Model Horse Shower's Journal* a real feel of community. Linda drew the covers, sketched in the margins, and typed every word. The covers, like this one from May 1978, also included photographic images. I read *TMHSJ* cover-to-cover on the day it arrived each month. There has never been anything like it. Linda still lives in Alabama, and enjoys attending BreyerFest each year.

Linda retyped everyone's ads, and devised a set of abbreviations for colors, breeds, brands, and show classes – such as "E.P." for "English pleasure" and "cos" for "costume" – that she used in the ads. The result was neatness, uniformity, clarity of meaning, and no wasted space. She fit the maximum amount of ads in her 56-page, monthly issues. Producing *MHSJ* was a full-time activity for her, and she sold it at a non-profit price. (Her parents subsidized her.) She produced it for 10 years. The last issue was January 1980.

Her readership then split itself among a handful of other model horse journals with fewer pages and fewer subscribers, and without the care that she had put into her publication. The Breyer company's publication was still only small, and anyway, the real action was in the publications produced by hobbyists. I subscribed to three of them.

In 1980, I entered my first photo show after joining Jo Maness' Made-in-Japan Club. For 16 years, I hardly missed entering a monthly show, and I judged annually. I still belong. I also joined Tina English's Hartland Lovers of America, and Laurel Orrin's Mini Model Horse Association, showed monthly for two+ years, and judged two or three shows for both clubs. Where I was living then, I did photography by crouching on a small balcony with the model on a chair in the doorway because there wasn't room for both of us.

In 1981, I entered my first in-person model horse show, held in Milwaukee by Jaci Bowman, a friend of Marney Walerius. Marney, of North Barrington, Illinois, had more or less invented in-person model horse showing ("live showing") about 1972. Certainly, she popularized it.

In February 1983, my first model horse book, on Hartland, came out. It was not the first book on a brand of model horses, but it was the second or third. The first book was actually not a book, but a stapled, multi-page list of Breyers published by Colleen Kenny (Waterhouse) in 1976. She later produced a stapled-edge Breyer book in 1990. The first real book was Cheryl Abelson's *Hagen-Renaker Handbook* in 1980. It included color photo-copied picture pages, and came in a plastic binder with the title and illustration stenciled on the binder cover. Her book showed how a book could be done without getting a publisher, who would need to be convinced that a venture would be profitable. With a subject like model horses, that could be difficult.

In 1983-1985, I was back in school, earning a master's degree in journalism. Since then, I've subscribed to many model horse journals that have come and gone, but were appreciated. They all did a lot to create and sustain a hobby community, as one of my journalism professors might say.

Most of my "career" since graduate school has consisted of voluntarily doing non-profit books and newsletters on model horses and Hartland models (in addition to some writing and editing for local history groups and some freelance research and organizing). I must be crazy! Well, this is how I have been spending my life. Everyone has to do something!

Around 1978, Linda Walter sold photocopies of two 1960s Hartland catalogs for 10 cents a page. I think that's when I found out that Hartland had originally been located in Wisconsin, less than 30 miles from my home. In 1974, I had seen an Illinois address on a Hartland catalog belonging to the downstairs tenant of my then-boyfriend. (Every relationship has a purpose!) The address belonged to what I later determined was the second Hartland company. When I called Illinois in 1974 to see if I could order models, they said production had ended the previous year. That was my first piece of research on Hartland.

I remember sort of daydreaming about Hartland in my parents' (our) living room in Milwaukee in about 1964. (The year can't be off by very much because we moved to Oak Creek in November 1966.) I'm sure I didn't know the company was so close by. I had no idea that I would later meet Hartland officials who were working there at that time – or maybe, somehow, I knew. In the 1990s, one of them asked me, "Where were you?" I replied, "I was 12!"

# CONCLUSION

The future is history repeating itself, only differently.

I started Hartland research in 1980 by locating and interviewing Hartland officials from the 1940s-1970s. When my first book came out in February 1983, Hartland horses had been off the market for 10 years. Within a year, Hartland horse production resumed.

I did 16 years of self-publishing Hartland books (1983-1998), and then *Hartland Horsemen* (1999) for Schiffer Publishing. When I turned in the work for *Hartland Horses and Dogs* to Schiffer in June 2000, it had been five and one-half years since Hartland horses were last made.

Two months later, I got a phone call from someone who said he was forming a new Hartland company. In October 2000, I was able to confirm that Hartland had, indeed, been sold to a new owner. The final proofing stage for *Hartland Horses and Dogs* came just in time to allow me to add a more upbeat ending to the chapter covering 60 years of Hartland history. Otherwise, it would have ended with, "After the long history of Hartland, it seems a shame."

*Hartland Horses and Dogs* was published in December 2000, and arrived in bookstores in January 2001. A month later, Hartland Collectibles, L.L.C., shipped its first new horse. Its last horse, to date, was in December 2007. Whether the brand is finished making horses or just napping (again), Hartland admirers have had seven bonus years.

*From left:* Bay, Champagne, and Buckskin foals from the 7" Tennessee Walker families shade from dark to light.

Now, there's this book. I'm not saying that a new book following a Hartland dry spell causes a flood, but when I think of Hartland, I have to smile.

*Children were awed. Old timers sighed. The crowd thronged. It was a splendid horse fair.*

A Bay 7" Tennessee Walker foal looks back at a fine time for Hartland.

# PUBLISHED WORKS BY GAIL FITCH ON HARTLAND MODELS, MODEL HORSES, OR REAL HORSES

## Individual articles, published in:

(Breyer's) *Just About Horses*, 1979 (Vol. 6, No. 1) – horse breeds quiz
*The Milwaukee Journal,* July 14, 1983 – explains the model horse hobby
*Wisconsin Sports Parade*, August 1983 – Grand prix jumping (real horses)
*The Model* [Horse] *Trading Post*, November 1994 – collecting psychology
*Made-in-Japan* [Horse] *Club*, September 1995 – artistry of Japan china horse figurines
*Toy Collector*, October 1995 – Hartland toys
*Model Horse Gazette,* March/April 1996 – Hartland horses
*Collecting Figures*, May 1996 – Hartland product history

## A series of articles, published in:

*TRR Pony Express*, 1994-1996, 11 articles – Hartland
*Horsing Around Magazine*, 2001-2004, 12 articles – Hartland, model horse, and equine art topics

## Books by Gail Fitch (self-published):

**Hartland Horses and Riders*, six editions (1983-1995) – model horses, horse-and-rider sets, and company history
**Horse Colors and Gaits,* three editions (1985-1990) – horse color and gaits as applied to models; illustrates many model horse brands, including Breyer
**Gift Horses from Japan* (1994, 1995) – Japan china horse figurines
**Hartland Horsemen and Gunfighters* (1998) – horse-and-rider sets, company history, and values
**Hartland Horses: New Model Horses Since 2000*, first edition (2010)

## Books by Gail Fitch, published by Schiffer Publishing LLC:

**Hartland Horsemen*, 1999 – horse-and-rider sets, with values
**Hartland Horses and Dogs, 2000* – model horses, with values, and company history

## Newsletters and miscellaneous published by GF:

^*Hartland Market*, 1994-1996 – 18 monthly issues totaling 194 pages; was 75% articles (on Hartland history and variations, mostly by GF)
^*Hartland Messenger*, 1999 – present – news update sheets; free with the books
Indexes for both Schiffer books, 2001; also called "Hartland Collecting Keys"
*Website, www.hartlandhorsemen.com, since July 2000 – news and more

Hartland Market

Vol. 1, No. 2 | Pages 9-16 | January 1995

**Steven Plans Taking Shape**

**December 28, 1994**—Steven Mfg. Co. and its Hartland division are in the process of deciding which model horses to make in 1995, according Liz Dalpe, customer service manager of Steven Mfg. After that decision is made, production could begin in February or March, she said. The company has recently undergone changes in management, but Liz Dalpe emphasized that it has *not* been sold. Also, some of the 1993-94 model horses are still in stock at Steven Mfg., she said. The new president of Steven Mfg. is Ty Bulkley, who has been in the toy manufacturing industry for 20 years, Liz Dalpe said. Former sales manager Don Light left in December for a position with Cadaco, a toy company in Chicago.

Above are two of Steven Mfg.'s finest current (1993-94) Hartlands, set #233, Bedouin Princess and Blessing. Dapple grey and black, they are of the "Jewel and Jade" molds sculptured by Kathleen Moody.

**Hartland Census Begins in January 1995**

*Hartland Market* is taking a census of 9" Tennessee Walkers and Large Western Champ Horses and Riders, with the results to be published in the February issue. Chances are that the models in our collections, taken as a whole, are fairly representative of the total population of models that exist or existed. The census will help suggest which models are more rare than others, which ones really exist, and which don't. If this seems like fun, we can do a census of different models each month. To participate, send your response by January 25th on models you currently own. Also, please see pages 10, 11, and 15.

**Hartland Collectables Sell Well**

**December 27, 1994**—Paola Groeber's December sale and auction of Hartland Collectables and other models was a success. Besides twelve 9" Saddlebreds in black with silver trim, 20 matching Weanling Foals, and four Pearl White Quarter Horses, only a handful of models remain.

Paola Groeber declined to disclose final bids, but said that she could answer individual questions on estimated values. She said that the proceeds will enable her to return to school to complete bachelor's and master's degrees in business administration.

The pearl white Quarter Horses resemble the 1960s pearl whites, rather than the usual late 1980s-1990 pearl whites, she said. Still available as of December 27 were two "charcoal" (deep brown with flaxen mane/tail) 9" Thoroughbreds, a special run chestnut Thoroughbred in styrene plastic, a 9" Grazing Arab Mare in sample color grey with black hooves, a 1984 Polo Pony in dark bay styrene, and the 1985 Christmas special Chestnut Arab Stallion and Weanling Foal set in styrene. For more information, write Paola Groeber, 1340 Morningside Dr., Mexico, Missouri 65265 or call (314) 581-1333.

From December 1994-May 1996, I published 18 monthly issues of *Hartland Market.* The newsletter included ads, but was mostly articles: updates for my Hartland book; news from the Hartland company, and the results of reader surveys I conducted to explore Hartland variations. *Hartland Market* had an unusual mix of readers: model horse collectors and Hartland horse-and-rider set collectors.

^*Hartland Model Equestrian News*, 2000-2007, 12 issues; renamed *Model Equestrian: Horse Collecting & Hartland News*, three issues through 2010; 350 pages in all; 90% articles, most by GF (on Hartland, model horse, and horse lovers' topics)

Key:
* Denotes works published largely or entirely in full color;
^ Denotes works with some color pages.